Lecture Notes in Electrical Engineering 1646

The book series *Lecture Notes in Electrical Engineering* (LNEE) publishes the latest developments in Electrical Engineering—quickly, informally and in high quality. While original research reported in proceedings and monographs has traditionally formed the core of LNEE, we also encourage authors to submit books devoted to supporting student education and professional training in the various fields and applications areas of electrical engineering. The series cover classical and emerging topics concerning:

- Communication Engineering, Information Theory and Networks
- Electronics Engineering and Microelectronics
- Signal, Image and Speech Processing
- Wireless and Mobile Communication
- Circuits and Systems
- Energy Systems, Power Electronics and Electrical Machines
- Electro-optical Engineering
- Instrumentation Engineering
- Avionics Engineering
- Control Systems
- Internet-of-Things and Cybersecurity
- Biomedical Devices, MEMS and NEMS

For general information about this book series, comments or suggestions, please contact leontina.dicecco@springer.com.

To submit a proposal or request further information, please contact the Publishing Editor in your country:

China

Jasmine Dou, Editor (jasmine.dou@springer.com)

India, Japan, Rest of Asia

Swati Meherishi, Editorial Director (Swati.Meherishi@springer.com)

Southeast Asia, Australia, New Zealand

Ramesh Nath Premnath, Editor (ramesh.premnath@springernature.com)

USA, Canada

Michael Luby, Senior Editor (michael.luby@springer.com)

All other Countries

Leontina Di Cecco, Senior Editor (leontina.dicecco@springer.com)

**** This series is indexed by EI Compendex and Scopus databases. ****

Hao-Shang Ma · Yu-Wei Chan
Editors

Proceedings of Innovative Computing 2026 Vol 4-Emerging Topics in Future Internet

Proceedings of IC 2026

Springer

Editors
Hao-Shang Ma
Department of Computer Science
and Information Engineering
National Taichung University of Science
and Technology
Taichung City, Taiwan

Yu-Wei Chan
Bachelor Deg Prog AI
National Taichung University of Science
and Technology
Taichung City, Taiwan

ISSN 1876-1100 ISSN 1876-1119 (electronic)
Lecture Notes in Electrical Engineering
ISBN 978-981-92-0778-7 ISBN 978-981-92-0779-4 (eBook)
https://doi.org/10.1007/978-981-92-0779-4

This Springer imprint is published by the registered company Springer Nature Singapore Pte Ltd.
The registered company address is: 152 Beach Road, #21-01/04 Gateway East, Singapore 189721, Singapore

Contents

Identity-Based Data Integrity Audit Scheme for Electronic Medical Records

Tao Wang[1(✉)] and Fushan Lu[2]

[1] School of Science, Kaili University, Kaili 556011, China
wangtao202509@163.com
[2] School of Microelectronics and Artificial Intelligence, Kaili University, Kaili 556011, China

Abstract. The medical system, as a fundamental aspect of public livelihood security, is progressively improving. Electronic medical records represent a significant advancement, with many countries worldwide dedicated to their development. Patients can utilize electronic medical devices for nationwide diagnosis and the storage of their medical records. This addresses the disparities in medical development across regions and significantly enhances the facilitation of medical consultations between doctors and patients. Consequently, ensuring data storage security and protecting patient identity privacy remain critical challenges that must be addressed in the realm of electronic medical records. This paper proposes an identity-based data integrity audit scheme specifically designed for electronic medical records. The study examines data protection and the security of user identities through the involvement of multiple parties. Additionally, a comprehensive proof of safety is provided. Finally, we compare several similar schemes, and the results demonstrate that our proposed scheme exhibits greater computational efficiency.

Keywords: cloud storage · data audit · identity-based · electronic medical records · privacy protection

1 Introduction

In the context of theoretical concepts such as the Internet of Everything, the Metaverse, and the Scientific and Technological Revolution1, the development of public infrastructure has accelerated significantly. Concurrently, medical security facilities, which serve as fundamental guarantees for the health and life of the populace, are also undergoing continuous improvements. Due to the independence of regional medical relationships, medical records between patients and hospitals can only be recognized on an individual basis. Nationwide certification of electronic medical records can not only optimize medical resources but also enhance the medical security of residents. In recent years, European and American countries have established comprehensive medical information systems within hospitals, while the United States, Japan, and other nations have invested significantly in the research and application of electronic medical records. Indiana University Hospital predicts the mortality rate of early-stage cancer patients by utilizing big data analysis of electronic medical records (EMR). The Hong Kong Hospital Authority

H.-S. Ma and Y.-W. Chan (Eds.): IC 2026, LNEE 1646, pp. 1–15, 2027.
https://doi.org/10.1007/978-981-92-0779-4_1

in China documents the patient's medical journey through patient cards, which include various elements such as CT examinations and clinical medications. In the United Kingdom, EMR is employed for maternity and pregnancy health assessments for expectant mothers. Concurrently, these countries are establishing specialized EMR laboratories dedicated to research. China has also undertaken extensive development of Electronic Medical Records (EMR). The introduction of national medical security cards has created a robust platform for EMR construction, facilitating the provision of national medical services and future enhancements. Concurrently, the People's Liberation Army General Hospital has engaged in the application and research of EMR2.

Electronic medical records are digital records created through the computer-based collection, storage, and access of medical information systems, and they possess legal validity. In comparison to traditional paper medical records maintained by individual hospitals, electronic records are easier to save, authenticate, and access. They offer users comprehensive data, as well as insights, predictions, and assessments of pathological causes, along with other relevant recommendations derived from big data analysis. In 2005, the United States released a over 500-page specification and certification standard for electronic medical records, which enhanced the national standards for electronic medical records. In 2005, the UK implemented electronic medical records (EMR) nationwide, which facilitated the selection of healthcare institutions, utilization of electronic medical records, registration services, and more. In 2003, Japan began implementing Electronic Medical Records (EMR) in certain regions, which provided legal advantages. The standard EMR application encompasses clinical data storage, clinical decision-making, controlled medical vocabulary, Internet of Things (IoT) order entry, prescription filling, and real-time clinical data applications. This system enables patients to utilize EMR for treatment in both outpatient and inpatient settings, with records being monitored by the government's third-party healthcare bureau. Additionally, these records are subject to review and delivery, holding legal significance in cases of disputes and medical certification, and are protected by law3.

The data integrity audit solution is an application designed for electronic medical records (EMR). Given the vast scale of national EMR systems, it is crucial to explore methods for ensuring data integrity and protecting patients' personal information from potential breaches. These aspects warrant thorough investigation and analysis. In the data integrity audit solution, storing Electronic Medical Records (EMR) as data in a cloud service provider (CSP) can significantly reduce local storage costs. Auditors are required to regularly audit the EMR data stored in the CSP to ensure data integrity. Additionally, when documenting medical records, the pseudonyms of patients and doctors are processed first to enhance the protection of patient personal information. By prioritizing data security and user privacy, this solution enables more effective and secure data storage.

2 Related Work

The proposal for the data integrity audit plan offers an effective solution for ensuring the security of data stored in cloud service providers (CSPs). By reducing user storage and computing costs, the plan introduces a third-party auditor (TPA) to verify data integrity within CSPs on behalf of proxy users. In 2003, Deswarte Y and colleagues4 first proposed a data integrity audit scheme utilizing the RSA cryptosystem. This article presents a conventional challenge-response protocol alongside an audit scheme based on the Diffie-Hellman protocol. Ateniese et al.5 proposed a provable data ownership model, S-PDP, which is based on the homomorphic tag algorithm. In comparison to other RSA-based data auditing schemes, this model is lightweight and significantly reduces server overhead. Xu et al.6 adapted the RSA algorithm proposed by Ateniese et al. for application within the blockchain context. They incorporated mechanisms for eliminating duplicate data audits 6 and for facilitating arbitrable data audits 7, thereby enhancing the overall usage scenario. In 2007, Juels et al.8 proposed a Proof of Retrievability (POR) audit scheme designed for ensuring the retrievability of large files. In this scheme, the original file is encrypted, and a set of randomly generated value-checking blocks, referred to as sentinels, are nested randomly within the file. However, this approach has the disadvantage of incurring a high computational cost. Hovav, Shacham et al.9 built upon the POR framework proposed by Juels et al.8 for the first time, providing a comprehensive security proof against any adversary in the strongest models, as established by Juels and Kaliski. The second solution is founded on a pseudo-random function (PRF), which permits only private verification and is secure within the standard model.

The issue of key exposure in cloud audit protocols has been highlighted10. Jia et al.11 first proposed an audit protocol that incorporates key exposure flexibility. This protocol is designed to detect any dishonest or malicious behavior, utilizing a binary tree structure 12 to efficiently update the user key. In 2008, Shah MA and colleagues13 proposed the implementation of a third-party auditor to arbitrate and ensure the integrity of data in the cloud, thereby enhancing user privacy protection. However, the involvement of inquisitive or dishonest third-party auditors also poses a risk of compromising cloud data security. The concept of public auditing was introduced 6, and Wang et al. 14 proposed a public auditing scheme aimed at protecting the privacy of shared data. By utilizing ring signatures to construct a homomorphic verifier, Oruta introduces a novel privacy protection mechanism that efficiently facilitates the simultaneous execution of multiple audit tasks. Wang et al. 15 proposed a scheme to enable scalable and efficient privacy-preserving public storage auditing in the cloud, utilizing random masking technology 16. This approach facilitates efficient multi-user audits within public audit environments. Additionally, Worku et al.17 introduced a secure and efficient public audit scheme for cloud storage privacy protection, employing ring signatures to maintain the confidentiality of shared users' tag identities during Third Party Auditor (TPA) audits. In a group user environment, W. Shen et al. 18 proposed a lightweight and privacy-preserving secure cloud auditing scheme for group users utilizing third-party media. The third-party auditor (TPA) performs time-consuming operations on behalf of users while protecting the original data during the upload and review stages. However, this scheme faces challenges from malicious users who may modify or delete data in the cloud. Yang et al.

19 proposed a public audit scheme for shared data in cloud storage that emphasizes privacy protection and traceability. However, this scheme faces challenges related to the dynamic nature of group users joining and leaving. Additionally, the threshold scheme introduced by Shamir et al. 20 utilizes secret sharing to distribute secrets among a group of participants, with each participant receiving a distinct secret value.

Waters et al. 21 first proposed an effective identity-based encryption (IBE) scheme in 2005, demonstrating its safety and efficacy within the standard model. In 2013, Wang H et al. 22 first proposed an identity-based remote data possession check (ID-RDPC) inspection protocol. In 2016, Zhang J et al. 23 introduced an efficient auditing mechanism for data within a multi-cloud storage environment, alongside an effective identity-based public auditing approach for outsourced data in cloud storage. This mechanism supports batch auditing in multi-cloud environments, reduces audit computation costs, and is more suitable for large-scale data storage. The identity-based data audit solution is particularly oriented towards practical applications. Additionally, Wang H et al. 24 proposed an identity-based agent data upload and remote data integrity check (ID-PUIC) protocol for the public cloud, which facilitates data uploads to the data owner through an agent, thereby accommodating complex network environments. Yong et al.25 proposed an identity-based remote data integrity checking mechanism aimed at ensuring robust data privacy protection for cloud storage. The Third Party Auditor (TPA) provides user privacy protection against untrusted third-party auditors. Additionally, by employing Zero-Knowledge Proofs 26, this approach can effectively counteract malicious servers, thereby enhancing its practical value. Shen W et al. 27 proposed a data audit scheme aimed at protecting sensitive information for group members within a shared cloud storage environment. This scheme ensures that while files are being shared, sensitive information remains concealed from other users, thereby safeguarding the privacy of user data.

There are studies exploring various aspects of remote data auditing, with this solution specifically concentrating on the application of electronic medical records. Deng et al. 28 introduced aggregated signatures in medical sensors, among other areas, while Shen W et al. 27 conducted related research on electronic medical records. This approach employs anonymous letters to manage patient information within electronic medical records, thereby ensuring anonymity and protecting identity concealment.

3 System Model and Related Concepts

3.1 Motivation and Contribution

This solution implements an auditing mechanism designed to safeguard the identities of both patients and doctors within electronic medical records. The contributions of this solution are as follows:

We propose a novel electronic medical record storage solution that generates a pseudonymous identity for each doctor and patient. This practical and succinct approach enhances the protection of user identities.

We propose a system model for an identity-based data audit scheme pertaining to electronic medical records, which emphasizes the security of private keys, the unforgeability of labels, and the anonymity of pseudo-identities.

Our solution demonstrates greater efficiency compared to other data audit solutions that utilize the same tags, and it also surpasses the performance of the previous solution.

3.2 Preliminary Knowledge

Relevant symbols and their definitions are provided in Table 1.

Table 1. Symbols and definitions

Symbol	Explain
p	A large prime number
G_1, G_2	P-order multiplication cyclic group
$g_1, g_2ς$	Generative elements in G_1
H_1, H_2, H_3, H_4	Four secure hash functions
Z_p^*	A set of positive integers less than p
e	A bilinear mapping
λ	An element in Z_q^*
$ID_i PID_i$	The patient's true identity and false identity
$ID_d PID_d$	The true and false identities of doctors
f_1	A pseudo-random function $\mathrm{f}_1 : \mathrm{Z}_\mathrm{q}^* \times \mathrm{Z}_\mathrm{q}^* \rightarrow \mathrm{Z}_\mathrm{q}^*$ $\mathrm{f}_1 : \mathrm{Z}_\mathrm{q}^* \times \mathrm{Z}_\mathrm{q}^* \rightarrow \mathrm{Z}_\mathrm{q}^* \mathrm{f}_2 : \mathrm{Z}_\mathrm{q}^* \times \mathrm{Z}_\mathrm{q}^* \rightarrow \mathrm{Z}_\mathrm{q}^*$ $\mathrm{f}_2 : \mathrm{Z}_\mathrm{q}^* \times \mathrm{Z}_\mathrm{q}^* \rightarrow \mathrm{Z}_\mathrm{q}^*$ $\mathrm{f}_2 : \mathrm{Z}_\mathrm{q}^* \times \mathrm{Z}_\mathrm{q}^* \rightarrow \mathrm{Z}_\mathrm{q}^*$
sk_{ID}	private key
$msk = g_2^\lambda$	System master key
F_{id}	Random value of file identifier
F	Original information file
m_i	The i-th data block of file F
ε	Replace elements in value G_1
$F' = (m_1{}', m_2{}', \ldots, m_n{}')$	Blocked files
σ_i	Signature label for file data block m_i
$chal$	Challenge information sent by TPA to the cloud, $chal = \{i, v_i\}_{i \in I}$
$proof$	Proof information generated by cloud

3.3 Model

The system model diagram is shown in Fig. 1.

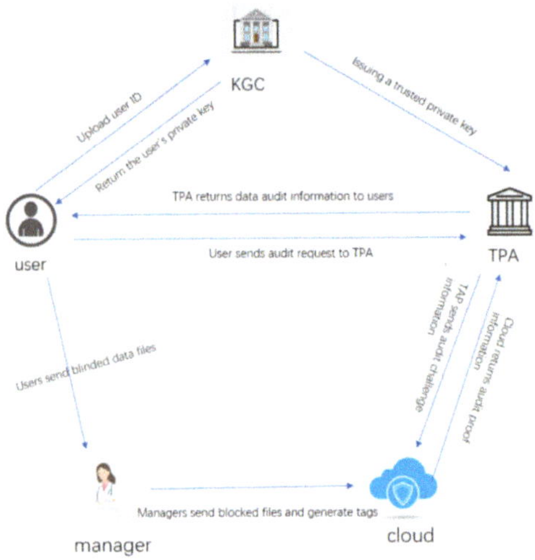

Fig. 1. System model diagram

Our system consists of four entities: a Private Key Generator (PKG), Patients (U), Managing Doctors (MD), a Cloud Service Provider (CSP), and a Third-Party Auditor (TPA). The PKG is a trusted authority that generates system parameters and issues private keys. Patients upload their EMR data to their managing doctor. The MD processes the data for privacy and uploads it to the CSP. The CSP, modeled as semi-honest, stores the data and responds to audit challenges. The TPA, whose trustworthiness may vary, performs integrity checks on behalf of users.

The identity-based data integrity audit scheme for electronic medical records comprises seven algorithms: **Parameter settings; Generate PID; Generate private key; Label algorithm; Shielding treatment; Generate proof; Verification proof.**

4 New Identity Based Data Auditing Scheme

4.1 Scheme Description

The specific algorithm for the new plan is as follows.

Parameter settings:

A bilinear map $e: G_1 \times G_1 \to G_2$ and a pseudo-random function $f_1: Z_p^* \times Z_p^* \to Z_p^*$, PKG selects two multiplicative cyclic groups, G_1 and G_2, both of prime order p, with g_1, g_2 serving as the generator of G_1.

PKG randomly selects $\lambda \in Z_p^*$ $\varsigma \in G_1$ and calculates $P_{pub} = g_1^{\lambda}$, $msk = g_2^{\lambda}$.

PKG selects 4 hash functions, $H_1: G_1 \to \{0,1\}^l$,$H_2: \{0,1\}^* \to G_1$,$H_3: \{0,1\}^* \to G_1$,$H_4: G_2 \to Z_p^*$.

PKG releases system parameter $PP = \left\{ \begin{matrix} g_1, g_2, e, p, G_1, G_2, f_1, \\ \varsigma, P_{pub}, H_1, H_2, H_3, H_4 \end{matrix} \right\}$ and saves the master key $msk = g_2^{\lambda}$.

PID generation:

The true identity of Patient U is ID_i, and PKG executes the following operations to create the pseudo-identity PID_i.

PKG randomly selects $c_i \in Z_p^*$ and calculates $PID_i = (PID_{i1}, PID_{i2})$, $PID_{i1} = g_1^{c_i} PID_{i2} = ID_i \oplus H_1\left(g_2^{\lambda} g_1^{c_i}\right)$.

PKG sends $PID_i = (PID_{i1}, PID_{i2})$ to patient U.

The true identity of Doctor D is ID_d, and PKG executes specific operations to create the pseudo-identity PID_d.

PKG randomly selects $c_d \in Z_p^*$ and calculates $PID_d = (PID_{d1}, PID_{d2})$,$PID_{d1} = g_1^{c_d}$, $PID_{d2} = ID_i \oplus H_1\left(g_2^{\lambda} g_1^{c_i}\right)$.

PKG sends $PID_d = (PID_{d1}, PID_{d2})$ to Doctor D.

Private key generation:

For the identity PID_i of patient U, the PKG arbitrarily selects the $x_{ID} \in Z_p^*$ and subsequently calculates the $Q = H_2(PID_i)$, $Y = Q^{x_{ID}}$, $sk_{ID} = \left(g_2^{\lambda} Y, g_1^{x_{ID}}\right)$.

For the identity PID_d of Doctor D, PKG arbitrarily selects $x_D \in Z_p^*$ and subsequently calculates $O = H_2(PID_D)$, $Z = O^{x_D}$, $sk_D = \left(g_2^{\lambda} Z, g_1^{x_D}\right)$.

Patient U confirms the accuracy of sk_{ID} using the following algorithm. $e\left(g_2^{\lambda} Y, g_1\right) = e\left(g_2, P_{pub}\right) e\left(Q, g_1^{x_{ID}}\right)$ Similarly, Doctor D verifies the correctness of SKD through the subsequent algorithm.

$$e\left(g_2^{\lambda} Z, g_1\right) = e\left(g_2, P_{pub}\right) e\left(O, g_1^{x_D}\right)$$

Tag generation:

For electronic medical record file $F = \{m_1, m_2, ..., m_n\}$, patient U/PID_i randomly selects $r \in Z_p^*$.

U performs tag processing on electronic medical record files and obtains the following $\sigma_i = g_2^{\lambda} Y (H_3(F_{id} i) \varsigma^{m_i})^r \Phi = \{\sigma_i\}_{1 \le i \le n}$ file tag set. The $F_{id} \in Z_p^*$ serves as a random value used for file identification.

The set U is defined as $\gamma_0 = F_{id} g_1^{x_{ID}} g_1^r$, from which the signature $\gamma = \gamma_0 Sig_{sk_{ID}}(\gamma_0)$ is calculated for $Sig_{sk_{ID}}(\gamma_0)$. This signature corresponds to the signing private key sk_{ID} applied to γ_0.

The user calculates the value $\varepsilon = \varsigma^r$ and subsequently sends both $\{F, \Phi, \gamma\}$ and ε to the managing doctor (MD), who then deletes these messages from local storage. When a patient or doctor wishes to retrieve and query electronic medical records, they can submit a request to the managing doctor. The MD then downloads and sends file F to the requesting user.

Shielding treatment:

The MD verifies the validity of the file signature tag γ by checking the signature $Sig_{sk_{ID}}(\gamma_0)$. If the validity is confirmed, the managing doctor proceeds to parse γ_0 in order to obtain the F_{id}, $g_1^{x_{ID}}$, g_1^r.

The management doctor MD passed the $e\left(\varsigma, g_1^r\right) = e(\varepsilon, g_1)$ test to verify the correctness of $\varepsilon = u^r$.

MD randomly selects a secret seed $k_1 \in Z_p^*$ as the input key for the pseudo-random functionf_1. Using the secret seed k_1, the management doctor calculates the blinding factor $\eta_i = f_{k_1}(i, F_{id})$ and subsequently computes $m_{i'} = m_i + \eta_i$. The resulting processed file is denoted as $F' = \{m_{1'}, m_{2'}, ..., m_{n'}\}$.

the management doctor MD calculates $\sigma_{i'} = \sigma_i(\varepsilon)^{m_{i'}-m_i}\sigma_{i'} = g_2^{\lambda}Y\left(H_3(F_{id}||i)\cdot\varsigma^{m_{i'}}\right)^r\Phi' = \{\sigma_{i'}\}_{1\le i\le n}$.

The MD uploads $\{\Phi', F'\}$ and γ to the cloud and subsequently sends γ to the TPA. After this, the MD deletes these messages from local storage. Meanwhile, the cloud service provider verifies the accuracy of label $\sigma_{i'}(i \in [1, n])$

$$e(\sigma_{i'}, g_1) = e(g_2, P_{pub})e\left(Q, g_1^{XID}\right)e\left(H_3(F_{id}||i)\cdot\varsigma^{m_{i'}}, g_1^r\right)$$

If the aforementioned formula holds true, the label is deemed valid; otherwise, the cloud service provider (CSP) will refuse to accept the data.

Proof generation:

The Trusted Third Party (TPA) verifies the validity of file tag γ. If file tag γ is determined to be valid, the TPA proceeds to parse γ_0 in order to obtain the file identifier name F_{id} and the experience value g_1^{rID}. Subsequently, the Generation Information Retrieval g_1^r generates the challenge information. Conversely, if file tag γ is found to be invalid, the TPA refrains from executing any further actions. TPA randomly selects a set I containing c elements, where I is a subset of $[1, n]$. Subsequently, TPA generates a random value $v_i \in Z_p^*.i \in I$ Finally, TPA produces challenge information $chal = \{i, v_i\}_{i\in I}$ and transmits it to the cloud.

After the cloud receives the $chal$, it randomly selects $t \in Z_p^*$ to compute $e\left(\varsigma^t, g_1^r\right) = B$. Subsequently, it calculates the linear combination of data blocks $M' = \sum_{i\in I} m_{i'}v_i$ and $M = t + \beta M'$, where $\beta = H_4(B)$. Following the computation of the aggregation label $\sigma = \prod_{i\in I}\sigma_{i'}^{v_i}$, the proof information $proof = \{M, \sigma, \varsigma^t\}$ is transmitted to the TPA.

Verification proof:

After TPA receives the file signature γ and the cloud's certification information $proof$, it calculates $e\left(\varsigma^t, g_1^r\right) = B$,$\beta = H_4(B)$ and verifies the accuracy of the file label using the following equation.

$$Be\left(\sigma^{\beta}, g_1\right) = e(g_2, P_{pub})^{\beta\sum_{i\in I}v_i}e\left(Q, g_1^{XID}\right)^{\beta\sum_{i\in I}v_i}e\left(H_3(F_{id}||i)^{\beta\sum_{i\in I}v_i}\cdot\varsigma^M, g_1^r\right)$$

If the equation holds true, the file in the cloud remains intact; otherwise, the file is considered corrupted.

4.2 Safety Analysis

In this section, we propose a scheme that ensures security with respect to correctness, pseudo-identity generation, and audit reliability.

Theorem 1: (Correctness of the Private Key) This scheme meets the following conditions necessary for the correctness of the private key: The correctness of the private key; Data label correctness; Audit integrity.

Prove: The correctness of the private key, the correctness of the data labels, and the correctness of the auditing phase have already been briefly addressed during the construction of the algorithm; hence, they are not restated here.

Theorem 2: (Audit Security) Calculating the Computational Diffie-Hellman Problem on a bilinear mapping is inherently difficult, and the method of label generation within this scheme is unforgeable. Specifically, for an adversary A or an untrusted cloud, if the data stored on the Cloud Service Provider (CSP) has been destroyed, it is theoretically computationally impossible for a forged certificate generated by the CSP to successfully pass the Third Party Auditor (TPA) verification process.

Prove: In the context of the security issues associated with this scheme, it is assumed that adversary A functions as an untrusted cloud provider, maliciously deleting user data to free up storage space. Additionally, A deceives users into believing they possess complete data and sends false certification information to the Third Party Auditor (TPA). If the Cloud Service Provider (CSP) can transmit proof information *proof* to successfully pass TPA verification despite the malicious deletion and destruction of data, it can engage in repeated interactions with adversary A by establishing a knowledge extractor to retrieve the challenged data block information.

Game 1: Challenger C and adversary A engage in an interactive query process. The adversary A queries the system parameter PP, prompting challenger C to execute the parameter setting algorithm to derive the system parameter PP and the master key msk. Challenger C then responds with the parameter PP. Subsequently, adversary A queries the private key sk_{ID}, leading challenger C to apply the private key generation algorithm to produce the sk_{ID} as a response; The adversary A queries the label, prompting the challenger C to execute the label generation algorithm and produce the output $\Phi = \{\sigma_i\}_{1 \le i \le n}$ as a response. During the challenge phase, the cloud service provider (CSP), acting as adversary A, receives the challenge information $chal = \{i, v_i\}_{i \in I}$ from the trusted third party auditor (TPA) or challenger C, and subsequently provides the proof information $\widehat{proof}$. In the forgery phase, adversary A successfully passes the TPA verification with negligible probability. If the forgery is successful, adversary A is declared the winner.

Game 2: The interaction between challenger C and adversary A mirrors that of game 1, with challenger C maintaining a query record concerning adversary A. Both challenger C and adversary A observe each instance of the process. If the label $\widehat{\sigma}$ and proof information $\widehat{proof}$ generated by adversary A successfully pass TPA verification, and differ from the original label $\Phi = \{\sigma_i\}_{1 \le i \le n}$ and complete proof information $proof = \left\{M, \sigma, \varsigma^t\right\}$, then adversary A's forged information successfully deceives the TPA, resulting in a victory for adversary A.

Game 3: The interactive query between challenger C and opponent A in this game mirrors that of game 2. However, challenger C continues to observe and retain a record of each query related to the proposed data integrity audit plan. In the event that one of the interactive queries reveals that the aggregated data block $\widehat{M}$ generated by adversary A does not match the original complete data information M, challenger C will declare the game terminated.

Through the execution of the aforementioned game and the accompanying mathematical analysis between the games, it follows that if the probability of a successful interaction between the adversary A and the simulator is negligible, then the constructed simulator is capable of solving the DL problem.

Theorem 3: (Security of Pseudo-Identity) Visitors to the data platform, as well as other participating users, are unable to access the true identities of patients and doctors. Furthermore, individual patients and doctors cannot view one another's true identities.

Game 4: Similar to Game 1, challenger C interacts with opponent A. The key difference is that challenger C maintains a list of records queried by the adversary. Adversary A succeeds if it can generate a fake identity.

Theorem 4: (Recognizability of Pseudo-Identities) The PKG is capable of identifying patients and users associated with pseudo-identities; specifically, only the PKG can restore the true identities of these pseudo-identities.

Proof: The Private Key Generator (PKG) generates the public parameter,and directly saves the master key $msk = g_2^{\lambda}$. The PKG arbitrarily selects a $c_i \in Z_p^*$. For the pseudo identity $PID_i = (PID_{i1}, PID_{i2})$, $PID_{i1} = g_1^{c_i}$, $PID_{i2} = ID_i \oplus H_1\left(g_2^{\lambda} g_1^{c_i}\right)$, we calculate $H_1\left(g_2^{\lambda} g_1^{c_i}\right)$. There is

$$PID_{i2} \oplus H_1\left(g_2^{\lambda} g_1^{c_i}\right) = ID_i \oplus H_1\left(g_2^{\lambda} g_1^{c_i}\right) \oplus H_1\left(g_2^{\lambda} g_1^{c_i}\right) \text{ Obtain;}$$

$$PID_{i2} \oplus H_1\left(g_2^{\lambda} g_1^{c_i}\right) = ID_i$$

Thus, the true identity of the signing user can be discerned from any given label.

Theorem 5: (Verifiability) The data processed by the management doctor (MD) is denoted as $F' = \{m_1\prime, m_2\prime, ..., m_n\prime\}$. Assuming that φ, $\varphi \in (1, n)$ data blocks are either damaged or tampered with in the untrusted cloud, and among these, c data blocks are specifically challenged, the probability of detecting any damage to the data blocks is at least $1 - \left(\frac{n-\varphi}{n}\right)^c$.

Game 5: Challenger C and Opponent A interact in a manner similar to that in Game 1, with Challenger C maintaining a record of the query concerning the opponent. If Adversary A is able to manipulate φ pieces of data with negligible probability while successfully passing verification, then the adversary is deemed to have succeeded.

Theorem 6: (For TPA privacy protection) When the Cloud Service Provider (CSP) responds to the Third Party Auditor (TPA) audit challenge, and subsequently sends the certification information proof to the TPA, it is important to note that the CSP cannot access the original data information $M' = \sum_{i \in I} m_i\prime v_i$.

Game 6: Challenger C interacts with opponent A in a manner analogous to game 1. Challenger C maintains a record of the query list pertaining to adversary A. When the Third Party Auditor (TPA), acting as adversary A, receives the proof of the Cloud Service Provider (CSP) response, the adversary succeeds if it can obtain the original data information $M' = \sum_{i \in I} m_i\prime v_i$ related to the challenge. Proof: Upon receiving the proof information $proof = \left\{M, \sigma, \varsigma^t\right\}$ from the CSP of challenger C, the adversary engages with the challenger and randomly selects β, $M \in Z_p^*$.In the context of the random oracle model, let $H(\cdot)$ be a machine for which it is essential to compute $\beta = H_4(B)$. For the adversary A, upon receiving the proof $proof = \left\{M, \sigma, \varsigma^t\right\}$ and subsequently calculating

$\beta = H_4(B)$, it becomes impossible to derive the original data information $M' = \sum_{i \in I} m_i{\prime} v_i$, resulting in the failure of adversary A.thereby ensuring the protection of user privacy.

Theorem 7: (For CSP data hiding) When managing uploads of doctor MD, both data and tags are concealed from the Cloud Service Provider (CSP), ensuring that user data remains private.

Proof: Before the MD uploads the data labels and data to the CSP, the MD first masks the data block, denoted as $m_i{\prime} = m_i + \eta_i$. Subsequently, the MD labels the masked data block as.

$$\sigma_i{\prime} = \sigma_i(\varepsilon)^{m_i{\prime} - m_i} = \sigma_i{\prime} = g_2^{\lambda} Y \left(H_3(F_{id} i) \varsigma^{m_i{\prime}}\right)^r$$

As a result, both the uploaded data and the corresponding data labels are masked, rendering it impossible for the CSP to access the original data information.

4.3 Efficiency Analysis

Our scheme is functionally compared with several related schemes [9, 15, 27, 29], as shown in Table 2. Notably, our solution is the only one that fulfills the requirements for anonymous letters, masked data, TPA data hiding, simplified certificate management, and public verification. We define the identical operations based on the comparison with the referenced paper.

Table 2. Function comparison of each plan

scheme	anonymous letter	Mask data	TPA data hiding	Simplified certificate	Public verification
scheme 9	NO	NO	NO	YES	YES
scheme 29	NO	YES	NO	YES	YES
scheme 15	NO	NO	YES	YES	YES
scheme 27	NO	YES	NO	YES	YES
Our scheme	YES	YES	YES	YES	YES

Table 3 illustrates that, in comparison to schemes [27] and [15], our proposed pseudo identity generation algorithm is unique, as no other scheme offers this feature, while the computational burden remains negligible. During the verification and proof phase, our computational efficiency significantly surpasses that of scheme [27], and our efficiency is comparable to that of scheme [15]. Furthermore, our algorithm provides a broader range of functions than scheme [15].

Table 3. Comparison of the computational cost of different schemes at each stage

stage	Our Scheme	Scheme [27]	Scheme [15]
Pseudo identity generation	$2Exp_{G_1} + Hash_{G_1} + Add_{Z_p^*}$		
Tag generation	$n(Hash_{G_1} + 2Exp_{G_1} + 2Mul_{G_1})$	$n(Hash_{G_1} + 2Exp_{G_1} + 2Mul_{G_1})$	$n(Hash_{G_1} + 2Exp_{G_1} + Mul_{G_1})$
Shielding treatment	$n\left(Sub_{Z_p^*} + Exp_{G_1} + Mul_{G_1}\right)$	$d_1 + d_2\left(Sub_{Z_p^*} + Exp_{G_1} + Mul_{G_1}\right)$	
Proof generation	$(c+1)\left(Mul_{Z_p^*} + Add_{Z_p^*} + Exp_{G_1}\right) + cMul_{G_T} + Hash_{G_2}$	$(d+1)\left(Mul_{G_1} + Add_{Z_p^*}\right) + cExp_{G_1} + cMul_{Z_p^*}$	$(c+1)\left(Mul_{Z_p^*} + Add_{Z_p^*}\right) + c(Mul_{G_1} + Exp_{G_1}) + Hash_{G_T}$
Verification proof	$(c+5)Exp_{G_1} + cMul_{G_1} + (c+1)Add_{Z_p^*} + (c+1)Hash_{G_1} + 4Pair$	$2Mul_{G_2} + 2(c-1)Add_{Z_p^*} + 2Exp_{G_2} + (l+c+1)Exp_{G_1} + (l+c)Mul_{G_1} + cHash_{G_1} + 4Pair$	$(c+3)Exp_{G_1} + (c+1)Mul_{G_1} + Mul_{G_T} + cHash_{G_1} + 2Pair$

This experiment utilizes the MIRACL library and implements the aforementioned operations on a personal computer (Lenovo, equipped with an Intel I5–3470 3.20GHz processor, 4GB of memory, and running Windows 7 operating system) (Table 4).

Table 4. Definition and execution time of related operations

Calculation	$Hash_{G_1}$	Exp_{G_1}	Mul_{G_1}	$Mul_{Z_p^*}$	$Add_{Z_p^*}$	$Hash_{Z_p^*}$	$Pair$	Mul_{G_2}	Exp_{G_2}
Duration(ms)	9.773	0.591	13.405	2.089	0.022	0.053	11.515	13.405	0.591

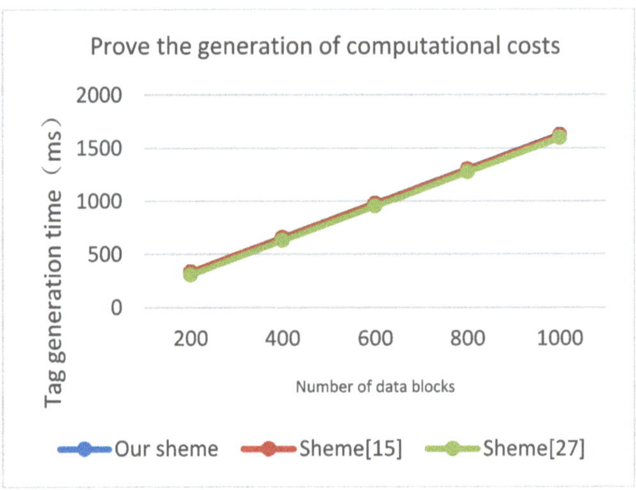

Fig. 2. Comparison of computational cost of label generation in different schemes

To effectively evaluate the performance of each stage, we establish the following fixed parameters. In this experiment, the total number of data blocks is set to 1000, with an interval of 200. During the masking stage, we apply masking to every 200 data blocks. As illustrated in Fig. 2, we compare Scheme [15] and Scheme [27] during the masking processing stage, demonstrating that our proposed scheme exhibits significantly greater computational efficiency.

Figure 2 illustrates that in the proof generation algorithm, 20 challenges are established for every 200 data blocks. Consequently, the computational cost of proof generation in our scheme is nearly equivalent to that of schemes [15] and [27]. However, under the same conditions, our scheme offers superior user identity protection features. In Fig. 3, we also maintain 20 challenges for every 200 data blocks while setting $l = 35$ in scheme [27] (where l denotes the number of users in scheme [27]). As the number of challenge data blocks increases, the computational cost of our scheme remains comparable to that of scheme [15], yet it is significantly higher than that of scheme [27]. Nevertheless, our scheme exhibits certain functional advantages.

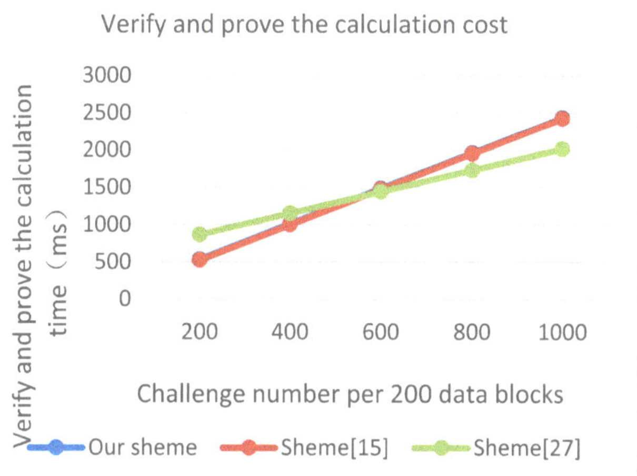

Fig. 3. Comparison of calculation costs for audit verification of different schemes

5 Summary

This article investigates and proposes an identity-based data integrity audit scheme for electronic medical records, which facilitates pseudo-identity generation and ensures data privacy protection for various users. This approach is significantly constructive for the advancement of remote electronic medical records. While safeguarding the true identities of patients and healthcare providers, it also encrypts medical record data, thereby achieving comprehensive protection of user privacy. Furthermore, the scheme allows for efficient and secure data integrity audits. Safety and efficiency analysis experiments demonstrate that this solution exhibits robust security and high efficiency.

Acknowledgement. Provincial Distinguished Teacher Zhang Heping Studio, Kaili University (MSGZS-SJ-2024003).

References

1. Tsinghua University's Metaverse Research Report: The Metaverse Spawns a New Wave of Social Interaction, Soul Becomes a Key Link in the Ecosystem Layout. People's Information [Accessed Date: 2021-11-11]
2. Baidu Baike entry for "Electronic Medical Record". https://baike.baidu.com/item/%E7%94%B5%E5%AD%90%E7%97%85%E5%8E%86#reference-[4]-18090-wrap.2022.11.11
3. Garets, D., Davis, M.: Electronic medical records vs. electronic health records: yes, there is a difference. Policy white paper. Chicago, HIMSS Analytics, pp. 1–14 (2006)
4. Deswarte, Y., Quisquater, J.J., Saïdane, A.: Remote integrity checking[C]//Working conference on integrity and internal control in information systems. Springer, Boston, MA, pp. 1–11 (2003)
5. Ateniese, G., Burns, R., Curtmola, R., et al.: Provable data possession at untrusted stores. Acm Conference on Computer & Communications Security. ACM (2007)
6. Xu, Y., Zhang, C., Wang, G., et al.: A Blockchain-enabled Deduplicatable Data Auditing Mechanism for Network Storage Services. IEEE Transactions on Emerging Topics in Computing (99), 1–1 (2020)
7. Xu, Y., Ren, J., Zhang, Y., et al.: Blockchain empowered Arbitrable data auditing scheme for network storage as a service. IEEE Trans. Serv. Comput. **13**(2), 289–300 (2020)
8. Juels, A.B.: Pors: proofs of retrievability for large files. ACM (2007)
9. Hovav, Shacham, Brent, et al.: Compact Proofs of Retrievability. J. Cryptol. J. Int. Ass. Cryptol. Res. (2013)
10. Yu, J., Ren, K., Wang, C., et al.: Enabling cloud storage auditing with key-exposure resistance. Info. Forensics and Secur. IEEE Trans. **10**(6), 1167–1179 (2015)
11. Jia, Y., Rong, H., Zhao, H., et al.: IRIBE: Intrusion-resilient identity-based encryption. Inf. Sci. **329**, 90–104 (2016)
12. Graham, R.L., Knuth, D.E., Patashnik, O.: Specific math: a foundation for computer science. Posts & telecom press, Beijing (2013)
13. Shah, M.A., Swaminathan, R., Baker, M.: Privacy-preserving audit and extraction of digital contents. Cryptology ePrint Archive (2008)
14. Wang, B., Li, B., Li, H.: Oruta: Privacy-preserving public auditing for shared data in the cloud. IEEE Trans. Cloud Comput. **2**(1), 43–56 (2014)
15. Wang, C., Chow, S., Qian, W., et al.: Privacy-preserving public auditing for secure cloud storage. IEEE Trans. Comput. (2013)
16. Wang, C., Wang, Q., Ren, K., et al.: Privacy-preserving public auditing for data storage security in cloud computing. 2010 Proceedings Ieee Infocom. Ieee, 1–9 (2010)
17. Worku, S.G., Xu, C., Zhao, J., et al.: Secure and efficient privacy-preserving public auditing scheme for cloud storage. Comput. Electr. Eng. **40**(5), 1703–1713 (2014)
18. Shen, W., Jia, Y., Hui, X., et al.: Light-weight and privacy-preserving secure cloud auditing scheme for group users via the third party medium. J. Netw. Comput. Appl. **82**, 56–64 (2017)
19. Yang, G., Jia, Y., Shen, W., et al.: Enabling public auditing for shared data in cloud storage supporting identity privacy and traceability. J. Sys. Softw. **113**(C), 130–139 (2015)
20. Shamir, A.: How to share a secret. Communications of the ACM (1979)
21. Waters, B.: Efficient identity-based encryption without random oracles. Annual International Conference on the Theory and Applications of Cryptographic Techniques, pp. 114–127. Springer, Berlin, Heidelberg (2005)
22. Wang, H., Wu, Q., Qin, B., et al.: Identity-based remote data possession checking in public clouds[J]. IET Inf. Secur. **8**(2), 114–121 (2014)
23. Zhang, J., Dong, Q.: Efficient ID-based public auditing for the outsourced data in cloud storage. Information Ences, 343–344, 1–14 (2016)

24. Wang, H., He, D., Tang, S.: Identity-based proxy-oriented data uploading and remote data integrity checking in public cloud. IEEE Trans. Inf. Forensics Secur. **11**(6), 1165–1176 (2016)
25. Yong, Y., Man, H., Ateniese, G., et al.: Identity-based remote data integrity checking with perfect data privacy preserving for cloud storage. IEEE Trans. Info. Forensics and Security (2017)
26. Ateniese, G., Faonio, A., Kamara, S.: Leakage-resilient identification schemes from zero-knowledge proofs of storage. Springer International Publishing (2015)
27. Shen, W., Qin, J., Yu, J., et al.: Enabling identity-based integrity auditing and data sharing with sensitive information hiding for secure cloud storage. IEEE Trans. Inf. Forensics Secur. **14**(2), 331–346 (2018)
28. Deng, L., Yang, Y., Gao, R.: Certificateless designated verifier anonymous aggregate signature scheme for healthcare wireless sensor networks. IEEE Internet Things J. **8**(11), 8897–8909 (2021)
29. Wang, B., Li, B., Li, H.: Panda: Public auditing for shared data with efficient user revocation in the cloud. IEEE Trans. Serv. Comput. **8**(1), 92–106 (2013)

LISA: A Label-Aware In-Place Scheduling Algorithm for High-Throughput Exact Subgraph Isomorphism

Lingyu Fang(✉)

Nanjing University of Finance and Economics, Nanjing 210023, China
fangly@163.com

Abstract. We investigate exact subgraph isomorphism (SGI) under a strict service-level agreement (SLA): at most, 60 s per target graph, with tens of thousands of query patterns per target. Our goal is to maximize the number of instances that finish within this fixed budget. We present LISA, a label-aware in-place scheduling algorithm that couples a label-degree lower bound (LD-LB) with cache-conscious candidate intersections, a connectivity-guided rarity-aware ordering with backup sequences, and a progress-aware round-robin scheduler. Compared to classical baselines (Ullmann, VF3), LISA reduces mean completion time by over 60% while raising the all-graphs full-match rate and aggregate hit rate. We provide a complete engineering blueprint—data structures, buffer reuse, time slicing, and determinism knobs—together with extensive experiments, ablations, and sensitivity studies. Our C++ implementation is lightweight and reproducible, making LISA practical for production SGI workloads.

Keywords: subgraph isomorphism · graph matching · pruning · vertex ordering · candidate intersections · scheduling · systems engineering

1 Introduction

Exact subgraph isomorphism underpins a wide range of real applications: substructure search in cheminformatics, pattern lookups in knowledge graphs, alert correlation in security analytics, and motif discovery in network science. These applications increasingly operate under an SLA that caps per-target runtime and favors throughput—finishing as many queries as possible in a given window—rather than unbounded, instance-by-instance completion. This regime shifts emphasis from raw worst-case complexity toward effective average-case reductions and robust scheduling under contention.

A modern SGI workload exhibits three practical traits. First, vertex labels are informative but sparse—the alphabet is small ($\leq$10) and non-uniform. Second, each target graph is accompanied by a large batch of patterns (10^4–5×10^4), including many easy instances and a long tail of hard ones. Third, deployment targets commodity single-socket CPUs, where cache locality and memory traffic dominate observed performance. Effective SGI under this workload calls for (i) extremely cheap early pruning, (ii) data layouts that exploit caches, and (iii) preemptive schedulers that avoid starvation.

H.-S. Ma and Y.-W. Chan (Eds.): IC 2026, LNEE 1646, pp. 16–24, 2027.
https://doi.org/10.1007/978-981-92-0779-4_2

We design LISA to satisfy these needs. The key idea is to make every hot-path operation trivially cheap—constant-time filters, in-place intersections, and O(1) state transitions—so that the scheduler can time-slice thousands of instances without paying heavy context-switch overheads. Our engineering goal is not just speed but predictability: similar instances should exhibit similar costs, and outliers should not derail batch progress.

Contributions—(1) A lightweight label-degree lower bound (LD-LB) that captures multi-label neighborhood constraints using precomputed per-vertex label counts. (2) Cache-friendly candidate generation via in-place intersections over label-bucketed adjacency, with a single reusable buffer per depth. (3) A connectivity-guided, rarity-aware ordering stabilized by multiple backups and a stall detector. (4) A progress-aware round-robin scheduler that improves fairness and coverage under a fixed time budget. (5) A complete, reproducible C + + implementation with deterministic options, accompanied by ablations and sensitivity analyses.

2 Related Work

Classic backtracking—Ullmann's matrix-based filtering [1]—established the canonical search-and-prune loop. VF2 [2] improved feasibility rules and state representation, while VF2 + + [3] refined ordering and propagation to cut branching factors on sparse graphs. VF3 [4] reorganized ordering and data access to better exploit modern caches and degree/label cues. Constraint-programming (CP) approaches (e.g., Glasgow [7]) leverage global constraints and domain reductions, often excelling on dense or highly regular instances at a higher constant factor. Database-oriented methods (QuickSI [5], TurboISO [6]) use decomposition and per-label indexing to seed small candidate sets before verification. Parallel and distributed systems push raw throughput on large clusters or GPUs but raise engineering complexity (partitioning, load balance, communication) and reproducibility concerns. Recent work on multi-pattern subgraph isomorphism explores dynamic scheduling strategies to improve throughput [8].

Our setting differs in two ways. First, the objective is coverage under a strict time budget—improving the number of completed instances—rather than minimizing time-to-first-solution for a single pattern. Second, labels are few yet discriminative, suggesting a particularly cheap filter based on label-wise neighbor counts. LISA revisits the sequential CPU design space, prioritizing constant-time checks, sequential memory access, and fine-grained preemption to raise batch throughput.

3 Background and Problem Setup

We consider undirected labeled graphs $G = (V, E)$ with label function $L : V \rightarrow \Lambda$, and connected query patterns $P = (V_P, E_P)$. A mapping $f : V_P \rightarrow V$ is a (non-induced) subgraph isomorphism if it is injective, preserves labels $(L(u) = L(f(u)))$ and edges $\left((u, u') \in E_P \Rightarrow (f(u), f(u')) \in E\right)$. We focus on small alphabets ($|\Lambda| \leq 10$), a regime common in chemistry and typed knowledge graphs.

Workload model—For each target, we process a batch of patterns. At $T = 60$s we report: (M1) average completion time over finished patterns; (M2) all-graphs full-match rate (fraction of targets for which all patterns finish); and (M3) overall hit rate (fraction of all individual pattern tasks finished). These metrics capture speed, robustness, and aggregate throughput under time pressure.

Challenge—Worst-case SGI is NP-complete, so no algorithm avoids exponential blowup in general. However, labeled sparsity, neighborhood structure, and careful engineering can reduce the effective branching factor by orders of magnitude on typical workloads.

4 Lisa Algorithm

A. Data Layout—We store labels as dense arrays and build per-label vertex lists V_ℓ. For each vertex v we precompute $\mathrm{cnt}(v, \ell) = |w \in N(v) : L(w) = \ell|$. Adjacency is kept sorted; we additionally build label buckets $N_\ell(v)$, enabling two-pointer intersections that touch only edges of the relevant label.

B. Rarity-Aware Connectivity Ordering—We rank pattern vertices by $\mathrm{score}(u) = \alpha \cdot \mathrm{rar}(L(u)) + (1 - \alpha) \cdot \deg(u)$, with $\mathrm{rar}(\ell) = 1/|V_\ell|$ computed on the target. The search maintains a connected frontier to activate edge constraints early. We precompute 3–5 alternative orders (vary α and tie-breakers). A stall detector switches to a backup if the current order produces no progress for several expansions.

C. Label-Degree Lower Bound (LD-LB)—For a frontier vertex u with multiset M(u) of required neighbor labels (accounting for matched neighbors), a candidate v is feasible only if $\mathrm{cnt}(v, \ell) \geq \mathrm{demand}_u(\ell)$ for all $\ell \in M(u)$. Because $\mathrm{cnt}(v, \ell)$ is precomputed, this check is constant time and runs before any set intersections.

D. In-Place Intersections with Injectivity—Given mapped neighbors $u' \to v'$, candidates for u must lie in $\left(\bigcap_{u' \in \Gamma_P(u) \cap \mathrm{dom}(f)} N_{L(u)}\left(f(u')\right)\right) \setminus \mathrm{Im}(f)$. . We intersect into a reusable buffer using two pointers; an occupancy array enforces injectivity (target vertices already used are skipped without probing hash tables). This design avoids allocations on the hot path.

E. O(1) DFS State—We keep forward/inverse maps and per-depth candidate vectors with indices. Assign/unassign updates two integers and one array cell. The tiny state enables preemption at millisecond granularity without expensive context save/restore.

F. Progress-Aware Round-Robin—Each pattern receives a quantum $\tau(p) = \tau 0 \cdot (1 + \beta \cdot |V_P|)$. . We prioritize (i) smaller patterns and (ii) deeper partial mappings. Preemption checkpoints the current depth and candidate index. A bounded-waiting property ensures no instance starves.

G. Determinism vs. Diversity—For reproducibility, LISA can fix RNG seeds and orderings. For coverage, we optionally inject light diversity (candidate shuffles or order cycling) when a pattern stalls.

H. Practical Defaults—Unless noted: $\alpha = 0.6$, $\tau 0 \approx 1.5$ ms, $\beta = 0.01$, , 4 backup orders. These values strike a balance across sparse targets and small label alphabets (Fig. 1).

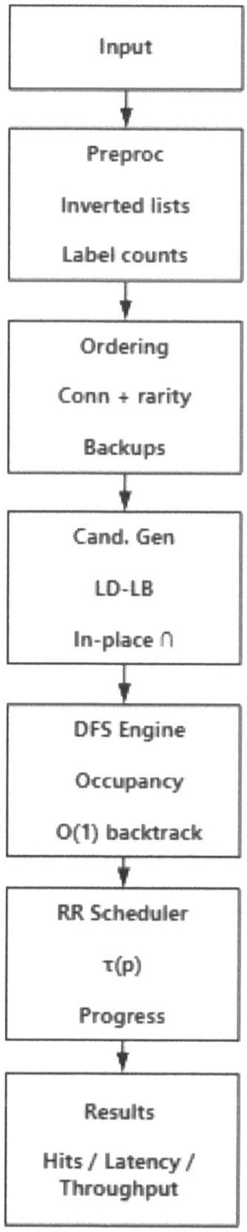

Fig. 1. Overall pipeline: preprocessing, ordering, LD-LB pruning, in-place intersections, DFS, and round-robin scheduling.

5 Correctness and Complexity

Soundness—LD-LB is a necessary feasibility condition because each unmatched labeled neighbor of u must map to a distinct labeled neighbor of v; insufficient $\mathrm{cnt}(v, \ell)$ violates edge preservation. Intersections over $N_\ell(v')$ retain exactly candidates that satisfy already-mapped edges. The occupancy array ensures injectivity at all depths.

Completeness—If a full embedding exists, no pruning step removes all of its consistent candidates: LD-LB is a lower bound, not a sufficient condition; intersections only remove vertices incompatible with mapped edges; injectivity does not forbid unused vertices.

Complexity—In the worst case, time is exponential in $|V_P|$. . Empirically, LD-LB coupled with early intersections collapses the branching factor. Per-step overhead is O(1) for checks and O(k) for intersections over small label buckets, with favorable cache behavior. Memory is $O(|V| + |E| + |\Lambda||V|)$ for precomputed counts and buckets, plus $O(|V_P|)$ per active depth.

6 Experimental Setup

Datasets—We evaluate on 500 targets spanning Erdős–Rényi and Barabási–Albert generators and real subsets from cheminformatics, KGs, and social graphs. Targets: $|V|$=500–2000, $|E|=3 \times 10^3$–1.2×10^4, $|\Lambda|\leq 10$. Patterns are connected with $|V_P|\leq 50$. We construct batches of 10^4–5×10^4 patterns per target to emulate production mix (many easy, some hard).Baselines & Environment—Ullmann and VF3 (non-induced). Intel Xeon E5–2690 v4 @ 2.60GHz, 128 GB DDR4 RAM, Ubuntu 20.04 LTS; GCC/Clang -O3; for fairness, all methods honor a 60s per-target budget. Seeds and binaries are archived for reproducibility.

Metrics—We report (M1) average completion time, (M2) all-graphs full-match rate, and (M3) overall hit rate at T = 60 s. We also break down percentiles (p50/p90/p99) and report memory peaks (Table 1).

Table 1. Dataset Summary

#Targets	Categories	|V|	|E|	|Λ|	Patterns/Target
500	ER/BA/Chem/KG/Social	500–2000	3×10^3–1.2×10^4	≤ 10	$10^4 - 5 \times 10^4$

7 Result and Analysis

Overall—LISA reduces mean completion time from 35.7s (VF3) to 11.2 s and more than doubles the fraction of targets where all patterns finish (38.0% → 61.0%). Aggregate hit rate rises from 61.8% to 80.1%. Time–completion curves show earlier gains, indicating faster pruning on easy and medium instances.

By Category—Gains persist across categories; BA graphs benefit most due to hub-driven constraints; chemical and KG targets profit from informative labels; social graphs remain hardest but still improve.

Runtime Breakdown—Profiler data attributes most cycles to two-pointer intersections and adjacency scans; LD-LB contributes modest cost with outsized pruning effect. Ablations confirm each component's role; scheduler is decisive for coverage under fixed budget (Figs. 2, 3, 4 and 5 and Table 2).

Table 2. Main Results (T = 60s, 500 Targets)

Method	Avg. Time (s)	Full-Match (%)	Hit Rate (%)
Ullmann	89.2	12.0	32.6
VF3	35.7	38.0	61.8
LISA	11.2	61.0	80.1

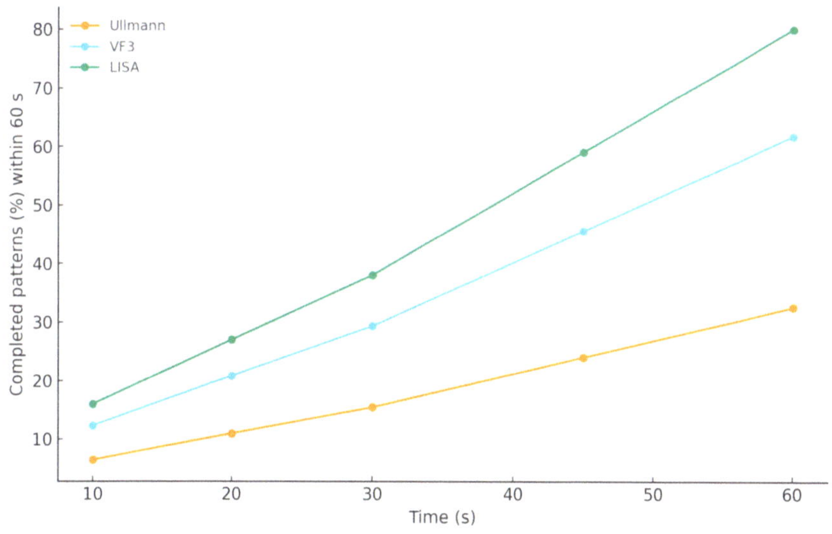

Fig. 2. Time–completion within 60s (Ullmann, VF3, LISA).

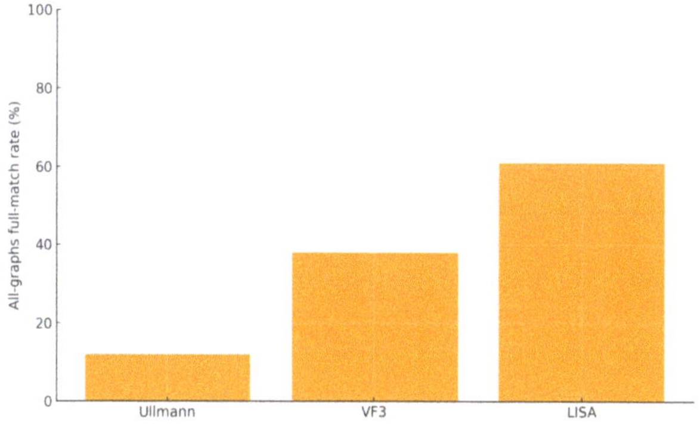

Fig. 3. All-graphs full-match rate by target.

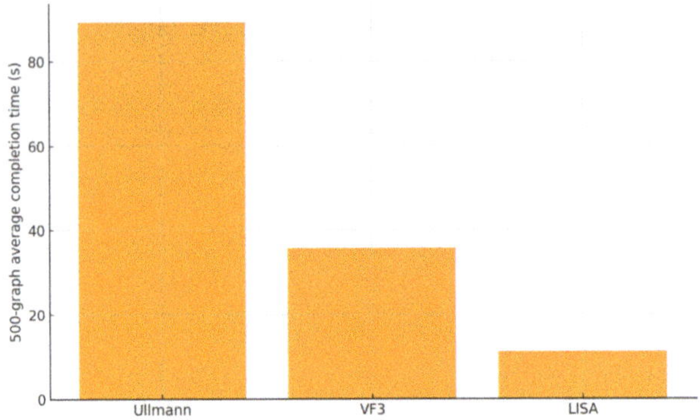

Fig. 4. Mean completion time by target.

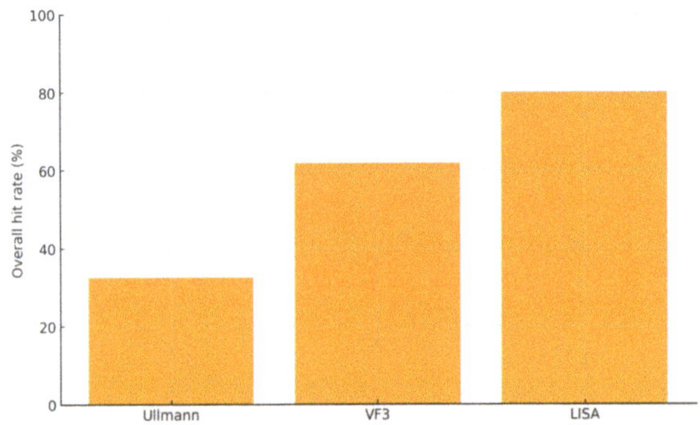

Fig. 5. Overall hit rate by target.

8 Ablation Studies

LD-LB Off—Average time increases sharply; candidate sets are larger, especially on BA/chemical targets where label neighborhoods are informative.

Intersections Off—Replacing label-bucketed intersections with naive scans causes a large constant-factor slowdown and more cache misses.

No Scheduler—A few hard patterns monopolize runtime; coverage collapses under the 60s cap.

No Order Diversity—Variance increases; some targets show long-tail stragglers that backups would have avoided (Table 3).

Table 3. Ablation Summary (Δ vs. Full LISA)

Removed	Δ Avg. Time	Δ Full-Match	Δ Hit Rate
LD-LB	+17.3 s	−2.7%	−1.8%
In-Place Intersections	+25.7 s	−1.9%	−0.9%

(*continued*)

Table 3. *(continued)*

Removed	Δ Avg. Time	Δ Full-Match	Δ Hit Rate
Round-Robin	+31.1 s	−23.4%	−22.5%
Order Diversity	+7.5 s	−0.8%	−0.3%

9 Parameter Sensitivity

α (rarity vs. degree)—Hit rate is stable for $\alpha \in [0.5, 0.7]$; too small undervalues labels in sparse regimes; too large over-constrains high-degree hubs.

$\tau 0$ (base quantum)—1–2ms balances preemption overhead and responsiveness; below 1 ms overheads grow, above 3 ms fairness degrades.

Backups—3–5 orders suffice; more yields diminishing returns unless the workload is adversarial (Table 4).

Table 4. Sensitivity Summary

Param	Range	Effect
α	0.4–0.8	Flat within 0.5–0.7; degrades outside
$\tau 0$ (ms)	0.5–4.0	U-shaped; best at 1–2
Backups	1–8	Best at 3–5; diminishing > 5

10 Engineering for Reproducibility

Fast I/O—A getchar()-based parser avoids locale/format overheads of iostreams and scanf, cutting input time by ~ 30% on large batches.

Label Buckets—neighborsByLabel[v][ℓ] stores only neighbors with label ℓ, minimizing spent cycles in intersections. Precomputing neighborLabelCount enables LD-LB as a constant-time gate before any set operations.

In-Place Intersections—Two-pointer merges write into a single reusable vector per depth; an occupancy array skips mapped vertices without map/set lookups.

Compact DFS State—Forward/inverse maps and per-depth candidate indices fit in L1/L2; assign/unassign are O(1). This keeps preemption cheap and predictable.

Build Flags—Release builds disable logging; typical flags include -O2/-O3, -pipe, -march = native. Deterministic runs fix RNG seeds and order tie-breakers when required by benchmarks (Table 5).

Table 5. Microbenchmarks

Optimization	On	Off	Δ Time
Fast I/O	4.2 s	6.0 s	– 30%
Label Buckets	11.2 s	18.1 s	– 38%
In-Place Intersections	11.2 s	36.9 s	– 70%
No Debug I/O	11.2 s	14.7 s	– 24%

11 Limitations and Threats to Validity

Graph Regimes—Dense graphs and extreme label imbalance diminish pruning benefits; completion times rise. Directed or multi-relational edges require extended feasibility checks and bucket structures.

Hardware Effects—Cache sizes and memory bandwidth influence constants; we report peaks but portability across microarchitectures warrants further study.

Dataset Coverage—While diverse, our mix cannot span all real workloads; future evaluations should include streaming updates and dynamic labels.

12 Conclusion

LISA demonstrates that careful coupling of label-aware lower bounds, in-place intersections, robust ordering, and fairness-oriented scheduling can dramatically raise SGI throughput under strict time budgets. Beyond raw speed, the design emphasizes predictability and ease of deployment. Future directions include learned order prediction, pre-filters for hopeless queries, and multi-core/GPU extensions.

References

1. Ullmann, J.R.: An Algorithm for Subgraph Isomorphism. J. ACM **23**(1), 31–42 (1976)
2. Cordella, L.P., Foggia, P., Sansone, C., Vento, M.: A (Sub)Graph Isomorphism Algorithm for Matching Large Graphs. IEEE TPAMI **26**(10), 1367–1372 (2004)
3. Jüttner, Madarasi, P.: VF2++—An Improved Subgraph Isomorphism Algorithm. Discrete Applied Math. **242**, 69–81 (2018)
4. Carletti, V., Foggia, P., Saggese, A., Vento, M.: Introducing VF3. GbRPR, LNCS **10310**, 128–139 (2017)
5. Shang, H., Zhang, Y., Lin, X., Yu, J.X.: Taming verification hardness: an efficient algorithm for testing subgraph Isomorphism. PVLDB **1**(1), 364–375 (2008)
6. Han, W.-S., Lee, J., Lee, J.-H.: TurboISO: Towards Ultrafast and Robust Subgraph Isomorphism Search in Large Graph Databases. In: SIGMOD, pp. 337–348 (2013)
7. McCreesh, C., Prosser, P., Trimble, J.: The Glasgow Subgraph Solver: Using Constraint Programming to Tackle Hard Subgraph Isomorphism Problem Variants, in Graph Transformation (ICGT 2020). LNCS **12150**, 316–324 (2020)
8. Ma, Z., et al.: Efficient Multi-Query Oriented Continuous Subgraph Matching. In: Proc. IEEE 40th International Conference on Data Engineering (ICDE), pp. 3230–3243 (2024)

Research on Public Participation Behavior of Environmental Governance Based on Big Data Analysis

Min Chen[1], Qicheng Yang[2(✉)], and Liangrui Zhang[1]

[1] School of Business Administration, Wuhan Business University, Wuhan 430056, China
[2] School of Public Administration, Zhongnan University of Economics and Law, Wuhan 430073, China
ttllttmm@163.com

Abstract. Environmental governance has long been a focal point for multiple stakeholders, with the public being a crucial participant. This study employs big data analysis methods, utilizing data collected and analyzed from online questionnaires, to examine the behavioral characteristics and influencing factors of public participation in environmental governance. Findings indicate that public willingness to participate in environmental governance does not align with actual participation behavior. While public willingness to participate is relatively high, the intention to engage in actual governance activities remains relatively low. Therefore, various measures should be implemented to enhance public intent to engage with environmental issues and increase communication and interaction among different stakeholders, thereby promoting the level and effectiveness of public participation in environmental governance. Building upon the study of public participation in environmental governance, this paper focuses on the public as its research subjects. It utilizes questionnaire surveys to understand the status of public participation in environmental governance, conducts empirical analysis on the data obtained from the questionnaires, and extracts relevant conclusions and enhancement strategies to increase the public's enthusiasm for participating in environmental governance.

Keywords: Big data Analysis · Environmental Governance · Public Participation · Behavior Survey

1 Introduction

The development of the national governance system is a strategic priority in the new era, with environmental governance serving as a crucial component. The 20th CPC National Congress Report states: "Advance the building of a Beautiful China, coordinate industrial restructuring, pollution control, ecological conservation, and climate change response, and promote ecological priority, resource conservation, intensive use, and green, low-carbon development." [1] Worsening pollution heightens public environmental awareness, with big data providing environmental information [2]. Government and enterprises should encourage public participation to address deficiencies in state and market systems [3].

H.-S. Ma and Y.-W. Chan (Eds.): IC 2026, LNEE 1646, pp. 25–36, 2027.
https://doi.org/10.1007/978-981-92-0779-4_3

Based on national conditions and existing research, this paper examines public participation in environmental governance. Through literature review, it establishes research methods, proposes hypotheses, and constructs models. Conclusion: Enhancing public environmental awareness and participation willingness is essential [4], leading to proposed measures. This research supports theoretical practice, provides a basis for collaborative governance, and advances the development of environmental governance systems.

2 Theoretical Basis

To analyze the behavioral logic behind such participation, this paper introduces two classic behavioral theories:

2.1 Planning Behavior Theory

TPB, proposed by Martin Fishbein and Icek Ajzen [5], explains the causes of specific behaviors and provides a foundation for behavior change, as shown in Fig. 1.

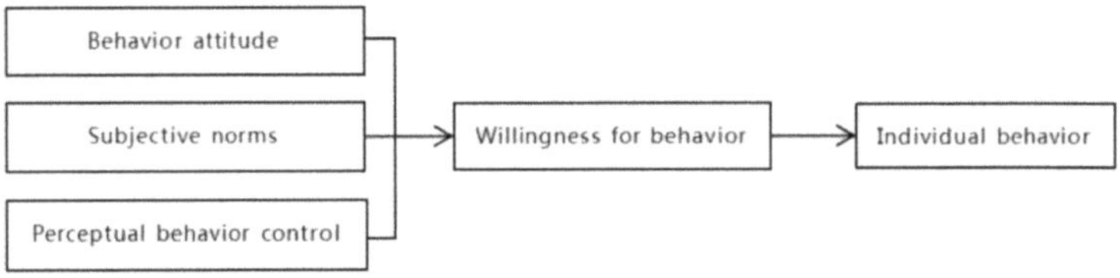

Fig. 1. TPB model

2.2 Normative Activation Theory

NAM was proposed by Schwartz in 1977 [6], indicating that moral norms influence behavioral change. Individuals activate behavioral norms under awareness of consequences and attribution of responsibility, thereby promoting their participation in environmental governance [7], as shown in Fig. 2.

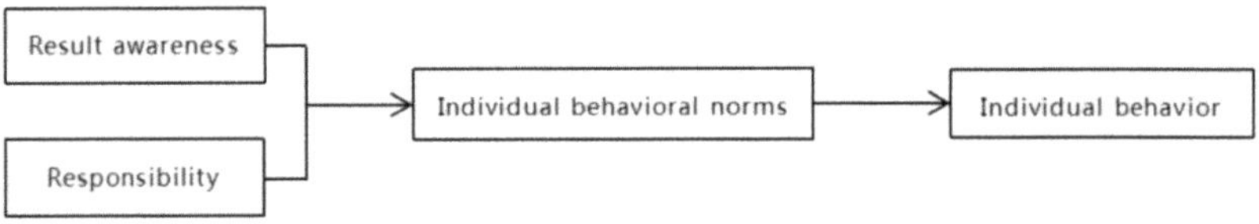

Fig. 2. NAM model

This paper builds an integrated model, and proposes a "social cognition—psychological control" research framework. Analysis of this model reveals that individuals' expectations regarding future outcomes are key factors influencing their behavior, with these expectations determining behavioral responses. This model facilitates a deeper understanding of behavioral motivations and optimizes the formulation of behavioral intervention strategies.

3 Current Status Analysis and Research Hypotheses

3.1 Current Status Analysis

Since the 1970s, both internationally and domestically, legal frameworks have been refined to legitimize public participation in environmental governance [8]. However, mechanisms require optimization and channels need expansion. Current public participation faces three challenges: high participation costs, insufficient government information disclosure, and high barriers for the public to access data [9], necessitating improvements to the information-sharing system; limited channels [10], and the lack of effective platforms diminishes outcomes, necessitating official platforms to boost engagement. Participation methods remain limited, often confined to post-decision stages, making it difficult for the public to influence decisions, thus requiring the expansion of diverse platforms. Environmental governance demands multi-stakeholder collaboration [11]. Figure 3 illustrates that multi-stakeholder co-governance involves entities collaborating toward shared objectives for mutual benefit. However, in practice, information asymmetry and self-interest can easily lead to divergent goals.

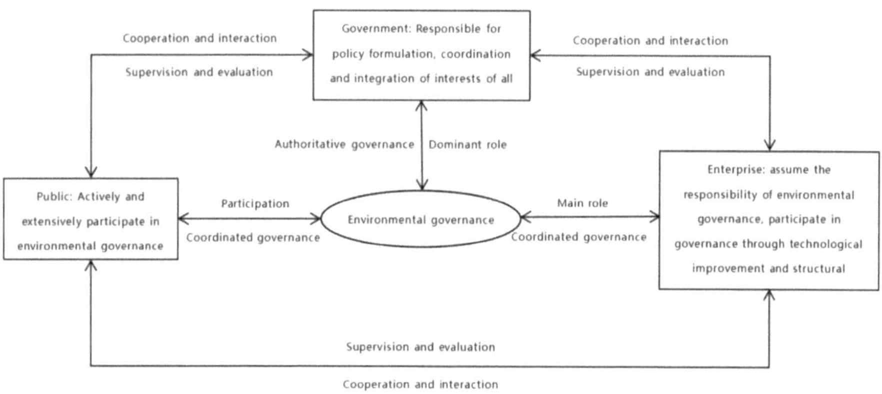

Fig. 3. Multi-dimensional co-governance model of environmental governance

3.2 Causal Relationship Analysis and Research Hypothesis

Causal Analysis.
Under the backdrop of technological and economic development, public willingness to participate in environmental governance has gradually increased [12]. This paper integrates two theories to construct a comprehensive system dynamics model encompassing eight variables.

Integrating TPB and NAM: Behavior can be predicted by assessing individual perspectives; based on this, hypotheses are proposed to construct a theoretical framework for public participation in environmental governance, as shown in Fig. 4.

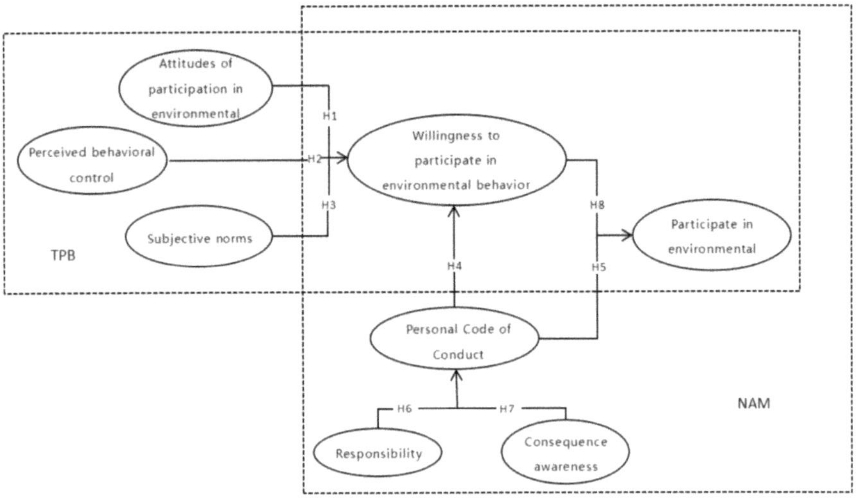

Fig. 4. Theoretical Research Framework

Research Hypothesis.

In TPB, public attitudes and behaviors influence participation willingness; individual participation is also affected by objective conditions. Government and corporate efforts to enhance public environmental capabilities can promote participation. These factors affect public behavior. Based on this, the following hypothesis is proposed.

H1: Participation attitude positively affects public willingness to participate

H2: Perceived behavioral control positively affects public willingness to participate

H3: Subjective norms positively affect public willingness to participate in environmental governance

In NAM, personal norms represent the sense of moral obligation when performing actions [13]; attribution of responsibility refers to the perceived impact of inaction by governments, enterprises, and the public, which influences individual participation in environmental governance; outcome awareness denotes the recognition of negative consequences resulting from neglecting environmentally harmful behaviors. Attribution of responsibility and outcome awareness can directly activate personal norms. Based on this, the following hypotheses are proposed.

H4: Personal behavioral norms positively affect public participation behavior

H5: The attribution of responsibility positively affects personal norms

H6: The awareness of consequence positively affects personal norms

H7: The norms of individual behavior positively affects the willingness to participate

H8: The willingness of public participation positively affects the behavior of participation

4 Model Design and Empirical Analysis

4.1 Research Samples and Data Sources

Surveys were distributed randomly using a seven-point Likert scale. The questionnaire comprised 49 items, with the first 40 measuring the scale and the last 9 collecting demographic information.

4.2 Representative Analysis of Research Data

Individual characteristics of the public may influence their environmental governance participation behavior either through specific categories or directly. This study randomly distributed 5,320 questionnaires. After excluding invalid questionnaires, 5,145 valid questionnaires were recovered, yielding a response rate of 96.71%.

The gender distribution among our interviewees is relatively even, though women outnumber men. There are differences in the channels through which people of different age groups obtain information about environmental governance. Based on the interview process, the respondents demonstrated a high level of education and were able to understand the content of the measurement indicators well. Occupationally, the majority were from the corporate sector, with a smaller proportion being government officials. Regarding income, most fell within the range of 5,001 to 10,000 yuan.

Overall, the sample in this study may exhibit certain biases. To understand the public's awareness of environmental governance activities, an additional question was included and analyzed as follows:

Public Attention to the Governance of the Surrounding Environment. From Table 1 above, we can know that the public's overall level of concern regarding environmental management in their surrounding areas is relatively high.

Table 1. Public attention

Way	Frequency	Percentage
Very concerned	455	8.84%
Pay more attention	1927	37.45%
focus on	2110	40.01%
Not paying much attention	609	11.84%
Don't pay attention	44	0.86%

The Frequency of the Public Joining Governance Activities in Daily Life. As shown in Table 2 above, it is necessary to organize more environmental activities and motivate the public to participate, working together to foster a positive environmental atmosphere.

Table 2. Frequency of public participation

Number of times	Frequency	Percentage
On average, 1–2 times per six months or less	408	7.93%
Average 1–2 times per quarter	1733	33.68%
Average 1–2 times per month	2063	40.10%
Average 1–2 times a week	824	16.02%
Average more than 3 times a week	117	2.27%

Number of Publicity Times of Environmental Governance Activities Carried Out by Surrounding Departments.
From Table 3 above, we can know that it is still necessary to strengthen publicity on environmental governance activities so that the public can better understand environmental governance publicity activities.

Table 3. Number of department promotions

Number of times	Frequency	Percentage
Once a week or more	726	14.11%
Once a month or more	2211	42.98%
Once or twice a year	1608	31.25%
Never	600	11.66%

Responsible entity for environmental governance.
From Table 4 above, we can know that environmental governance cannot rely solely on a single entity but necessitates coordinated participation from multiple stakeholders.

Table 4. Responsible entity

Number of times	Frequency	Percentage
public	702	13.64%
government	619	12.03%
enterprise	1339	26.03%
All of the above	2485	48.30%

4.3 Empirical Analysis

Reliability Analysis.
In this paper, the Cronbach coefficient method is used to test the internal consistency of the questionnaire. From Table 5, we can see that the total Cronbach's discharge coefficient of the entire questionnaire is 0.967, and the Cronbach's discharge coefficient of the eight dimensions is above 0.9. This shows that the internal consistency coefficient of this questionnaire is high, the questionnaire test results have high reliability and stability.

Table 5. Reliability Analysis

Dimension	Cronbach's α	Number of questions	Total scale reliability
AT	0.916	5	0.967
PB	0.919	5	
SN	0.917	5	
PN	0.917	5	
AC	0.928	5	
AR	0.931	5	
EW	0.920	5	
EC	0.924	5	

Validity Analysis.
(1) Exploring factor analysis

Table 6. KMO and Bartlett's test

KMO		0.958
Bartlett's Sphericity test	χ^2	175460.137
	df	780
	P	0.000

The questionnaire was subjected to KMO test and Bartlett spherical test. From Tables 6, it can be seen that the results show that its KMO value is 0.958 > 0.7, indicating that the data can be used for exploratory factor analysis.

The results of exploratory factor analysis show that the reliability content of the research project can be effectively extracted. In addition, the corresponding factor loadings of items within the same dimension are all greater than 0.6, demonstrating the good structural validity of the data.

(2) Confirmatory factor analysis

Table 7. Distinguishing validity

	AT	PB	SN	PN	AC	AR	EW	EC
AT	0.832							
PB	0.512	0.839						
SN	0.499	0.519	0.832					
PN	0.512	0.501	0.493	0.837				
AC	0.490	0.494	0.501	0.525	0.852			
AR	0.478	0.493	0.500	0.410	0.510	0.857		
EW	0.508	0.527	0.402	0.514	0.517	0.502	0.837	
EC	0.518	0.530	0.530	0.519	0.337	0.439	0.531	0.843

Note: The diagonal number is the square root value of AVE

As shown in Table 7 above, the diagonal values for each dimension exceed the corresponding correlation coefficients, indicating that the scale data possesses good discriminant validity.

(3) Correlation Analysis

Table 8. Correlation Analysis

	AT	PB	SN	PN	AC	AR	EW	EC
AT	1							
PB	0.512**	1						
SN	0.499**	0.519**	1					
PN	0.512**	0.501**	0.493**	1				
AC	0.490**	0.494**	0.501**	0.525**	1			
AR	0.478**	0.493**	0.500**	0.410**	0.510**	1		
EW	0.508**	0.527**	0.402**	0.514**	0.517**	0.502**	1	
EC	0.518**	0.530**	0.530**	0.519**	0.337**	0.439**	0.531**	1

* $P < 0.05$ ** $P < 0.01$

As shown in Table 8, correlation analysis and Pearson correlation coefficients indicate significant positive correlations among all dimensions, with correlation coefficients ranging from 0.3 to 0.6. These correlations are suitable for subsequent structural equation modeling.

(4) Structural equations

To investigate the influence relationships among various dimensions, a structural equation model was constructed with 8 latent dependent variables and 40 observed variables. e1–e40 represent the error terms of observed variables, while e41–e43 denote the residual terms of latent dependent variables. The fitted structural equation model is illustrated in Fig. 5 below.

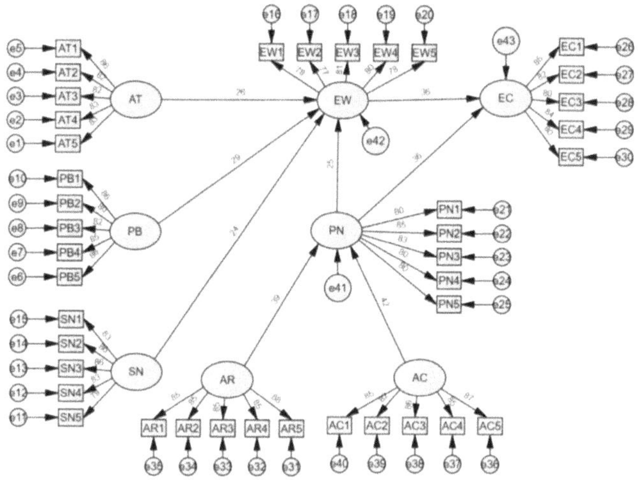

Fig. 5. Fitting coefficient diagram of structural equation

4.4 Structural Model Fit Test

First, the model fit was analyzed. The $\chi 2/df < 3$, and most other indicators passed the test values, indicating that the model fits well (Table 9).

Table 9. Model fitting index

Index	χ^2	*df*	*p*	χ^2/df	GFI	RMSEA	RMR	CFI	TLI	NNFI
Acceptable	-	-	> 0.05	< 3	> 0.9	< 0.10	< 0.05	> 0.9	> 0.9	> 0.9
Value	1130.019	723	0.000	1.563	0.799	0.052	0.038	0.916	0.909	0.909

4.5 Structural Model Test Results

As shown in Table 10: AT (0.257), PB (0.293), SN (0.238), and PN (0.249) all exhibit significant standardized path coefficients for EW ($P < 0.05$). This indicates that behavioral attitude, perceived behavioral control, subjective norm, and personal norm positively influence participation willingness (H1-H4 are established), meaning that the public's positive attitude, strong sense of control, and heightened sense of responsibility correlate with stronger participation willingness. AR (0.387) and AC (0.423) exhibit significant path coefficients for PN ($P < 0.05$), confirming that responsibility attribution and outcome consciousness positively influence personal norms (H5-H6 established). Participation by friends and relatives can drive public behavior; The path coefficients of EW (0.361) and PN (0.361) on EC are significant ($P < 0.05$). Personal norms and participation willingness positively influence participation behavior (H7-H8 established), meaning those with a strong sense of responsibility and strong willingness are more likely to participate.

Table 10. Model path analysis results

path			Estimate	standard estimate	S.E	C.R	P
EW	←	AT	0.226	0.257	0.012	18.284	< 0.001**
EW	←	PB	0.268	0.293	0.013	20.754	< 0.001**
EW	←	SN	0.238	0.238	0.014	16.964	< 0.001**
EW	←	PN	0.262	0.249	0.015	17.654	< 0.001**
EC	←	PN	0.484	0.361	0.019	25.161	< 0.001**
PN	←	AR	0.253	0.387	0.009	28.295	< 0.001**
PN	←	AC	0.362	0.423	0.012	30.449	< 0.001**
EC	←	EW	0.461	0.361	0.019	24.74	< 0.001**

Public participation in environmental governance embodies responsibility and honor, driving environmental practices, strengthening governmental commitment and societal oversight, and optimizing legislative governance. It contributes to ecological conservation, economic development, and improved livelihoods, making its enhancement an essential measure for building a better environment. This paper examines public participation from the perspective of citizens, integrating two theoretical frameworks to construct an analytical model. The findings are as follows:

Public willingness and behavior toward environmental governance are positively driven by individual environmental attitudes, surrounding environmental influences, personal norms, and a clear understanding of accountability and consequences. Simultaneously, governments, businesses, and the public must clearly define their respective governance responsibilities, establish a reasonable accountability framework, and through multi-stakeholder collaboration, transform willingness into tangible actions.

5 Optimize Public Participation in Environmental Governance Behavior Strategies

The preceding section employed SPSS 25.0 and AMOS 25.0 to construct structural equation models, testing hypotheses regarding public participation willingness, exploring participation mechanisms, and analyzing relevant influencing factors. Proposals were formulated from three dimensions—government, enterprises, and the public—to achieve sustainable development and clarify participation outcomes and directions.

5.1 Improve the Social Governance Structure and Subject Relationship

Proposing solutions from three dimensions. Government: Lead governance by enforcing oversight, establishing environmental policies, disclosing information, and streamlining reporting channels; leverage platforms to share data and regulations, conduct environmental advocacy, promote green consumption guidelines, incentivize public participation through rewards, and strengthen government-business-public collaboration; Enterprises: Promote public engagement, establish industry standards, and enhance brand value through technological innovation; Public: Participate in environmental activities, practice green lifestyles, and monitor environmental policies.

5.2 Promote Enterprise Green Supply Chain Technology Innovation

Enterprises should innovate green technologies at the source, optimize supply chains, select environmentally friendly production methods, reduce waste and pollution, and establish sustainable models. Promote green consumption through gifts and reward points. Simultaneously, establish feedback mechanisms to encourage consumer participation in monitoring. Additionally, set standards, drive green innovation, organize team training, build risk management systems, and align environmental and business objectives.

5.3 Optimize Public Participation Behavior Patterns

Public-led participation can enhance environmental awareness, but it faces challenges such as weak coordination, high costs, and difficulty in quantifying outcomes, necessitating management oversight. Spontaneous organizations can precisely address environmental issues. Improvements can be made by refining participation channels, enhancing platform engagement, and establishing incentive mechanisms; strengthening public responsibility and trust; and facilitating communication among government, enterprises, and citizens. Public participation not only fosters a positive social atmosphere but also enhances the effectiveness of initiatives. While public participation remains limited, sustained government guidance and appropriate mechanisms can boost public enthusiasm, thereby improving governance outcomes.

Acknowledgement. This paper is supported by "2022 Hubei Provincial Social Science Fund Pre-Funding Project for Major Philosophical and Social Science Research in Higher Education Institutions" (22ZD140).

References

1. Wei, Z.: Twenty Top Reports draw up a new blueprint for China entering a new journey of modernization. China Food **20**, 10–17 (2022)
2. Sijie, L.: Using big data for "environmental intelligence." People's Forum **23**, 74–75 (2018)
3. Ren, Z.: Public participation in third-party governance of environmental pollution: interest game and legal system improvement. J. Henan Univ. (Social Science Edition) **62**(04), 52–57+153 (2022)
4. Jinyu, S., Yuqi, W., Zheng, Z.: Factors influencing public participation in urban water environment governance behavior - based on survey data from Beijing, Shanghai and Guangzhou. Resource Science **43**(11), 2289–2302 (2021)
5. Hill, R.J., Fishbein, M., Ajzen, I.: Belief, attitude, intention and behavior, an introduction to theory and research. Contemp. Sociol. **6**(2), 244–245 (1977)
6. Schwartz, S.H.: Normative influences on altruism. Adv. Experimental Social Psychol. **10**(C), 221–279 (1977)

7. Yuexing, Z., Sijing, W.: Building a "empowerment-identification-cooperation" mechanism for public participation in environmental governance—a research based on planned behavior theory. Fujian Forum (Humanities and Social Sciences Edition) **10**, 169–176 (2017)
8. Zhang, G., Deng, N., Guan, X., Cheng, S., Bao, H.: Impact of public environmental supervision behavior and public environmental participation policies on the efficiency of industrial pollution control——Empirical analysis based on Chinese provincial panel data. China Populat.·Resour. Environ. **29**(01), 144–151 (2019)
9. Jianbin, G.: Public participation in environmental governance: logical review and practical selection. Jiangxi Soc. Sci. **41**(04), 84–95 (2021)
10. Yan, Z., Li, L.: Environmental governance of grassroots local governments and environmental satisfaction of rural residents—a questionnaire survey based on farmers in the L River Basin of Hunan Province. Shanghai Economics **01**, 72–84 (2020)
11. Guo, J., Xu, Y.: Logic, path and effect of public participation in environmental governance. Resource Science **42**(07), 1372–1383 (2020)
12. Yu, Z.: Institutional efficiency, corporate environmental investment and corporate value. University of International Business and Economics (2018)
13. Yuan, G., Fang, H.: Research on pro-environmental behavior in rural residents' domestic waste treatment - Taking Wenliu Town, Henan Province as an example. Sci. Technol. Indus. **22**(06), 18–25 (2022)

A Deep Learning-Based Model for Identifying Abnormal Access Behavior and Security Protection in Low-Carbon Park Networks

Chengxi Yin[1,2(✉)], Haolin Wang[1,2], Yun Zhao[1,2], Yuxin Lu[1,2], Ziwen Cai[1,2], and Xipeng Liu[1,2]

[1] Electric Power Research Institute of CSG, Guangzhou, China
ycx_csg@163.com
[2] Guangdong Provincial Key Laboratory of Intelligent Measurement and Advanced Metering of Power Grid, Guangzhou, China

Abstract. The network system of low-carbon parks is responsible for various important tasks, including energy control, equipment status monitoring, and user service support. However, abnormal access behavior caused by network attacks not only directly affects the normal operation of business, but may also lead to a sudden increase in energy consumption during emergency response, further causing an increase in carbon emissions. The commonly used anomaly detection methods based on rules or statistical features are often difficult to handle the rapid changes and complex situations of network behavior patterns. Therefore, this article proposes a multimodal anomaly recognition and dynamic protection framework that combines deep learning mechanisms. On the basis of summarizing existing research results, this article carefully analyzes the shortcomings of existing methods and focuses on designing lightweight feature extraction modules, adaptive threshold adjustment strategies, and low resource consumption protection mechanisms. This model not only maintains high detection accuracy, but also fully considers the overall energy efficiency of the system, providing a feasible technical route for network security management in low-carbon parks.

Keywords: Deep learning · Low carbon park · Abnormal network access behavior · Security protection mode

1 Introduction

With the continuous promotion of the "dual carbon" strategy, the low-carbon park network, as an important infrastructure, is responsible for tasks such as energy regulation and equipment status monitoring. This type of system transmits various data in daily operation, such as information collected by sensors, records generated by user terminals, and interactive content between different management platforms. At the same time, the system also needs to complete real-time control tasks, so it faces various security threats: external attacks, including malicious intrusion and large-scale distributed denial of service attacks, often occur; Internal risks include personnel unauthorized access to

H.-S. Ma and Y.-W. Chan (Eds.): IC 2026, LNEE 1646, pp. 37–45, 2027.
https://doi.org/10.1007/978-981-92-0779-4_4

sensitive information and other operational behaviors. In addition, the system also needs to deal with highly covert infiltration activities, such as low-frequency and slow type network attacks. Once abnormal access occurs, it may not only affect the normal scheduling process of energy, but also activate emergency response mechanisms, which often result in additional energy consumption and increase carbon emissions levels. The commonly used detection methods are no longer able to adapt to the increasingly complex and diverse network behavior characteristics. Therefore, it is necessary to adopt deep learning methods to establish a dynamic security protection model that can accurately identify anomalies and has low resource consumption.

2 Literature Review

2.1 Limitations of Traditional Anomaly Detection Methods

Early research mainly focused on two technical directions, the first using rule matching, such as Snort intrusion detection system; The second method relies on statistical analysis, such as threshold judgment or mean variance models. Rule matching requires manual updating of the rule library, making it difficult to keep up with changes in new attacks [1]. Statistical methods assume that normal behavior follows a certain probability distribution, such as a Poisson process. But the actual network traffic is dynamic and often has multiple patterns, such as different access habits on weekdays and holidays. This characteristic results in relatively high false positives and false negatives in traditional methods [2]. There are also studies indicating that in low-carbon parks, the false alarm rate of using fixed thresholds for traffic monitoring may reach 15% to 20% due to seasonal performance fluctuations.

2.2 Improvement and Shortcomings of Machine Learning Methods

At the beginning of the 21st century, machine learning began to be used in this field. Supervised learning methods such as support vector machines and random forests, as well as unsupervised learning methods such as K-means clustering and isolated forests, have been applied. But supervised learning requires a large amount of labeled data, which means experts need to accurately distinguish normal and abnormal data in advance. In low-carbon parks, there are already very few attack samples, especially unknown attack data that is even more difficult to obtain. Unsupervised learning, although not requiring labeled data, is not effective in handling high-dimensional features such as protocol type, access time, and packet size: the dimensionality reduction is not ideal, and it is easily affected by noisy data. Comparative studies have shown that in the industrial Internet of Things environment, the F1 score of random forests is only 0.78, but practical use generally requires a score of 0.9 or above, indicating that the performance of existing methods is not good enough [4].

2.3 Progress and Adaptation Issues of Deep Learning Methods

In recent years, deep learning technology has gradually become a research hotspot due to its ability to automatically extract features. In this field, some typical models have been widely used. For example, recurrent neural networks are suitable for processing time-series traffic data, but these models suffer from the problem of vanishing gradients. In addition, when the sequence is very long, the computational resources required for training will significantly increase, leading to a decrease in efficiency.

Convolutional neural networks perform well in capturing local features, such as identifying patterns in specific fields of packet headers, but their ability to model overall temporal dependencies is relatively limited. Autoencoders generally detect abnormal behavior by reconstructing errors, with small reconstruction errors for normal samples and large errors for abnormal samples. However, in complex application environments, autoencoders with relatively simple structures often have difficulty effectively distinguishing anomalies.

In addition, most existing deep learning methods have a common problem: they mainly focus on anomaly detection, but do not fully integrate protection mechanisms. Many models only output abnormal judgment results and lack effective coordination with network security policies, making it difficult to form a complete closed-loop protection system.

Meanwhile, some models have a large number of parameters, and when deployed on edge devices, it significantly increases the energy consumption and computational burden of the devices, which is an important challenge that needs to be addressed in practical deployment [5].

3 Model Design

3.1 Overall Architecture

The model is divided into four levels, processing data step by step and performing protective operations:

Data collection layer: installed on the park gateway and core switch, the collection module is responsible for collecting raw network traffic data, such as NetFlow logs; Simultaneously obtaining device access records, including HTTP requests and SSH login behavior; In addition, it also receives metadata from sensors, including access time, device type, and geographic location information.

Preprocessing layer: This layer processes different types of data separately. Firstly, perform data cleaning to remove duplicate and invalid records; Then perform normalization operations, such as scaling numerical features using the Min Max method; Finally, the multimodal data is encoded separately, for example, the TF-IDF method is used for text-based URLs, while sliding windows are used to extract statistical features for time-series traffic data [6].

Detection layer: As the intelligent judgment core of the system, this layer includes a deep learning module composed of a feature extraction network and an anomaly classifier.

Protection layer: responsible for linking with security equipment in the park, including firewalls and software defined network controllers, dynamically adjusting protection measures based on detection results, such as blocking high-risk source IPs or imposing bandwidth restrictions on abnormal devices.

3.2 Multimodal Feature Extraction and Fusion

In the low-carbon park scenario, abnormal behavior usually involves the interconnection between multiple information dimensions, such as an IP address attempting to log in multiple times in a short period of time and repeatedly accessing sensitive interfaces. Therefore, it is necessary to conduct a unified analysis by integrating temporal features, statistical features, and semantic features. Therefore, this article constructs the following types of feature combinations (Table 1):

Table 1. Multimodal Feature Combination

Feature type	Specific indicators	Instructions
Temporal features	Access frequency (times/minute), standard deviation of request interval, session duration, and burst traffic (peak/mean)	Reflecting the dynamic patterns of user behavior, abnormal access often manifests as high-frequency or low-frequency mutations
Statistical features	Request success rate (percentage of 200 status codes), average/variance of packet size, distribution of protocol types	Abnormal requests may be accompanied by high failure rates (such as brute force cracking) or unconventional protocols (such as non-standard port HTTP)
Semantic features	Login username/password complexity, URL path depth, parameter anomalies (such as SQL injection keywords)	Identify content-based attacks (such as credential explosion, web vulnerability exploitation)
Device association features	Consistency of device types with the same MAC address, matching degree between IP and geolocation (such as accessing the intranet from an IP outside the campus)	Detecting forged identities or cross regional abnormal connections

For multimodal features, this study uses a combination of modal encoding and attention fusion for processing. Local relationships are extracted using a one-dimensional convolutional network for temporal features, with a convolution kernel size of 3 and 32 filters; Statistical features and semantic features are mapped to the same dimension through fully connected layers. Afterwards, a multi head self attention mechanism is added to automatically learn the weights of each modality, and finally fused into a 64 dimensional comprehensive vector. This approach not only reduces the common information duplication in traditional feature concatenation, but also significantly reduces the computational complexity. Compared with long short-term memory networks (LSTM), the total number of model parameters is reduced by about 40% [7].

3.3 Lightweight Deep Detection Model

To achieve a balance between detection accuracy and energy efficiency, this paper designs a lightweight autoencoder (LAE) as the main classifier for anomaly detection. This structure has two advantages: firstly, it adopts unsupervised training, does not require a large amount of labeled data, and is suitable for practical situations where attack samples are scarce in low-carbon parks; Secondly, through structural optimization, the number of parameters is significantly reduced, and the entire model size is only about 120KB, which can be deployed in embedded gateway devices based on ARM Cortex-A53 architecture [8].

Model Structure Improvement.
In terms of model structure, the lightweight autoencoder proposed in this paper adopts the basic form of encoding decoding, with the main goal of identifying abnormal behavior by reconstructing input features. Specifically, during model training, efforts are made to minimize the gap between the original input and the reconstructed output: normal samples typically have smaller reconstruction errors, while abnormal samples have larger errors. This process can be represented by the following mathematical formula.

$$L = \frac{1}{n}\sum_{i=1}^{n}(x_i - \hat{x}_i)^2 \tag{1}$$

Among them, x_i represents the value of the original input feature in the i-th dimension, and $\hat{x}_i$ is the corresponding reconstructed value output by the decoder. N represents the total dimensionality of the input features, with $n = 64$ in this article. This loss function guides model training through backpropagation algorithm, helping the encoder effectively learn key features from normal behavior data.

Improvement Point Explanation.
Reducing the dimension of the intermediate representation from 64 to 16 improves the model's ability to extract key features. At the same time, a residual connection mechanism has been added to the encoder structure, specifically for the mapping process from 32 dimensions to 16 dimensions. This connection adds the input x_{32} to the output x_{16}, and its mathematical expression is:

$$x_{16}^{out} = x_{16} + x_{32} \tag{2}$$

Among them, x_{16} is the input of the 16 dimensional layer, x_{32} is the output of the 32 dimensional layer, and gradient disappearance is alleviated by residual.

Quantitative perception training: Simulate 8-bit integer quantization (converted during actual deployment) to reduce inference energy consumption.

After the encoder of this model inputs 64 features, it sequentially passes through two fully connected layers. The first layer has 32 nodes and uses the ReLU activation function; The second layer compresses the features to 16 dimensions and also uses the ReLU function. The decoder adopts a symmetrical design, first restoring the 16 dimensional features to 32 dimensions, and finally reconstructing them into a 64 dimensional output. The output layer uses the Sigmoid activation function to keep the reconstructed values within a valid range. When identifying anomalies, the mean square error between the original input and the reconstructed result is calculated to evaluate and determine whether there is an anomaly.

In terms of model optimization, the following three improvements have been made: firstly, compressing the bottleneck layer from 64 dimensions to 16 dimensions to help the model learn more representative features and improve generalization ability; Secondly, residual connections are added to the encoder to sum the input and output of the layer from 32 to 16 dimensions, which not only reduces gradient decay but also accelerates training convergence; Thirdly, a quantization aware mechanism is introduced during training to simulate an 8-bit integer computing environment, allowing the model to be converted to INT8 format during deployment. Experiments have shown that these optimizations increase inference speed by approximately 2.1 times, while reducing energy consumption by up to 35%.

3.4 Adaptive Threshold Optimization Strategy

In traditional autoencoder models, anomaly detection generally uses a fixed threshold, such as adding three times the standard deviation to the average reconstruction error as the basis for judgment. However, in application scenarios such as low-carbon parks, the normal energy usage pattern will constantly change with seasons and time, for example, an increase in air conditioning usage in summer will result in a significant increase in flow. If the fixed threshold method is still used at this time, it may lead to erroneous alarms [9].

To address this issue, this article proposes a dynamic threshold setting method that can automatically adjust based on statistical features in sliding windows. Specifically, assuming there are *k* normal samples in the current window, their corresponding reconstruction error sequences are $\{L_1, L_2.., L_K\}$. First, calculate the average value μ_w and standard deviation σ_w of the error within the window. Based on these statistical results, the dynamic threshold T can be calculated using the following formula:

$$T = \mu_w + k\sigma_w \tag{3}$$

In the model parameter settings, add a variable *k* called "sensitivity coefficient" with a value between 2.5 and 3.0, and the specific size will be selected based on the results of the validation set. If the reconstruction error *L* new of the current sample is greater than the set threshold *T*, it is considered abnormal.

Historical normal traffic data is divided by hours, for example, using weekdays from 9am to 6pm as a time period. At each time interval, calculate the average mean square error (MSE) of all samples, μ and the standard deviation σ. During real-time detection, the system automatically selects the nearest time interval based on the current time and dynamically sets the anomaly detection threshold *T*, using the formula $T = \mu + k\ \sigma$. Among them, the value of parameter *k* is also between 2.5 and 3.0, and is adjusted through the validation set.

For occasional abnormal behaviors, such as newly discovered attack methods, this article proposes a method for accumulating suspicion. Specifically, if the MSE of the sample exceeds the temporary threshold T '(set as $\mu + 2\ \sigma$) three times in a row, the system will issue an alarm. This method can effectively reduce false alarms caused by single fluctuations.

3.5 Design of Dynamic Protection Mechanism

In the dynamic protection mechanism, once an exception is detected, the system will cooperate with the SDN controller and firewall in the park through the northbound interface to perform hierarchical processing: for low-risk exceptions, such as an IP frequently accessing non critical interfaces in a short period of time, the request frequency of that IP will be limited, for example, not exceeding 10 times per second, and a log will be recorded [10]; For medium risk anomalies, such as brute force login or attempting to access the management configuration interface, the IP will be temporarily banned, usually for 30 min, and the administrator will be notified; For high-risk anomalies, such as discovering data theft or control instructions being modified, the system will immediately block the relevant traffic by setting DROP rules through the firewall and isolating the affected devices. This means that the SDN controller will close the corresponding port and activate emergency plans, such as switching to a backup control channel.

4 Experimental Verification and Result Analysis

4.1 Experimental Environment and Dataset

The experimental server is equipped with Intel i7-11800H processor and NVIDIA T4 graphics card, and also uses Raspberry Pi 4B as the simulation deployment platform for the edge gateway. In terms of data, the study uses two types of datasets: the first type is synthetic attack samples generated based on the MITRE ATT&CK framework, including 12 typical attack types; The second type is real desensitization data from a national low-carbon demonstration park, including three months of network traffic records and system access logs, of which 5% are labeled abnormal samples. In terms of model comparison, four typical methods were selected: traditional algorithm Isolation Forest (IForest), supervised learning method Random Forest (RF), classical time series model LSTM-AE, and the LA-E model proposed in this paper.

4.2 Evaluation Indicators

In terms of detection performance, accuracy, precision, recall, and F1 value are used; In terms of energy efficiency indicators, it mainly depends on the time consumption of single inference, model storage volume, and edge deployment power consumption; In terms of protection effectiveness, it is measured by the false blocking rate (the proportion of normal requests that are intercepted incorrectly) and the success rate of attack blocking.

4.3 Experimental Results

Comparison of Detection Accuracy.
As shown in Table 2, in the evaluation of detection accuracy, the model proposed in this

paper achieved an F1 value of 0.93 on the test set, which is significantly higher than the comparison methods (IForest is 0.71, RF is 0.82). This advantage mainly comes from two aspects: on the one hand, multimodal feature fusion effectively extracts the correlation features between behaviors; On the other hand, the lightweight encoder structure enhances the model's generalization ability to unknown attack patterns.

Table 2. Comparison of Accuracy Testing Results

Model	Accuracy	Precision	Recall	F1-score	Parameter quantity (KB)
IForest	0.78	0.65	0.70	0.71	< 1
RF (Supervision)	0.85	0.80	0.83	0.82	210
LSTM-AE	0.89	0.85	0.87	0.88	320
LA-E model	0.94	0.92	0.95	0.93	120

Energy Efficiency and Real-Time Performance.
The test results on Raspberry Pi 4B edge devices are shown in Table 3, indicating that this model has greatly improved energy efficiency. In addition, the use of dynamic thresholding method further improved the detection performance, reducing the false alarm rate from 12% when the threshold was fixed to 6%, while the false alarm rate was also below 3%, maintaining a very low false alarm rate.

Table 3. Comparison of Energy Efficiency and Real time Performance

Model	Single inference time consumption	Power consumption
LA-E model	12ms	180mW
LSTM-AE	45ms	320mW

Verification of Protective Effect.
To evaluate the actual effectiveness of the protection mechanism, this study used simulated attack methods for verification. Specifically, using the Metasploit framework to launch brute force attacks on the target SSH service. The test results show that the protection system can effectively intercept about 98% of high-risk intrusion behaviors, while only 0.5% of normal access requests are erroneously blocked. Further analysis reveals that the phenomenon of false blocking is mostly caused by transient fluctuations in the network, which lead to sudden changes in the MSE index. Overall, the system has significantly improved the security protection strength while ensuring business continuity, meeting the high reliability and low interference requirements for network security proposed by low-carbon parks.

5 Conclusion

This article proposes a deep learning based multimodal detection and dynamic response architecture for abnormal access behavior in low-carbon park networks. This method combines multi-source features to construct a lightweight autoencoder network, and incorporates an adaptive threshold adjustment mechanism to achieve efficient anomaly recognition performance. Specifically, the model achieves an F1 value of 0.93 and the inference time in edge devices is controlled within 12 ms. In addition, the system has achieved collaborative linkage with software defined network controllers and firewalls, establishing a complete protection system. Experimental data shows that the proposed model outperforms traditional solutions in terms of detection accuracy, energy efficiency performance, and actual deployment capability, and can provide effective support for network security management and carbon emission control goals in low-carbon parks.

References

1. Kala, E.S.M.: Critical role of cyber security in global economy. Open J. Safety Sci. Technol. **13**(4), 231–248 (2023)
2. Priyadarshini, R., Barik, R.K.: A deep learning based intelligent framework to mitigate DDoS attack in fog environment. J. King Saud Univ.-Comp. Info. Sci. **34**(3), 825–831 (2022)
3. Sauber-Cole, R., Khoshgoftaar, T.M.: The use of generative adversarial networks to alleviate class imbalance in tabular data: a survey. J. Big Data **9**(1), 98 (2022)
4. Blázquez-García, A., Conde, A., Mori, U., et al.: A review on outlier/anomaly detection in time series data. ACM Comput. Surv. (CSUR) **54**(3), 1–33 (2021)
5. Gomez-Rosero, S., Capretz, M.A.M.: Anomaly detection in time-series data using evolutionary neural architecture search with non-differentiable functions. Appl. Soft Comput. **155**, 111442 (2024)
6. Wang, H., Miller, D.J., Kesidis, G.: Anomaly detection of adversarial examples using classconditional generative adversarial networks. Comput. Secur. **124**, 102956 (2023)
7. Koren, O., Koren, M., Peretz, O.: A procedure for anomaly detection and analysis. Eng. Appl. Artif. Intell. **117**, 105503 (2023)
8. Atli, B.G., Miche, Y., Kalliola, A., et al.: Anomaly-based intrusion detection using extreme learning machine and aggregation of network traffic statistics in probability space. Cogn. Comput. **10**(5), 848–863 (2018)
9. Chen, J., Qi, X., Chen, L., et al.: Quantum-inspired ant lion optimized hybrid k-means for cluster analysis and intrusion detection. Knowl.-Based Syst. **203**, 106167 (2020)
10. Attou, H., Guezzaz, A., Benkirane, S., et al.: Cloud-based intrusion detection approach using machine learning techniques. Big Data Mining and Analytics **6**(3), 311–320 (2023)

Rainstorm Disaster Zoning in the Eastern End of the Hexi Corridor Based on AHP and Geographic Information System

Yunfeng Duan(✉)

College of Meteorology, Lanzhou University of Resources and Environment, Lanzhou 730021, China
okay2010@163.com

Abstract. Based on the theories of natural disaster systems and disaster risk assessment, and combining with the relevant theories of rainstorm disaster risk assessment, this paper selects the eastern end of the Hexi Corridor as the study area. Using the observational data from conventional meteorological stations, historical disaster situation data, and basic geographic information data in this region, it explores and studies the technical methods for rainstorm disaster risk zoning in arid and semi-arid areas. By analyzing four indicators—hazardness of disaster-causing factors, sensitivity of disaster-bearing environment, vulnerability of carriers, and disaster prevention and mitigation capabilities—a rainstorm disaster risk assessment model was established by using the Analytic Hierarchy Process (AHP) and weighted comprehensive evaluation. Additionally, the Geographic Information System (GIS) technology is employed to obtain the disaster risk zoning of this region. The results show that the high-risk areas of rainstorm disasters in the eastern end of the Hexi Corridor are mainly distributed in the main urban areas of each county/district, the southern part of Liangzhou District, and most areas of Gulang County. Tianzhu County is located in the medium-to-high risk zone, while other regions have relatively low risks, with Minqin County having the lowest risk. Overall, the high-risk areas for rainstorms are concentrated in the southern mountainous areas with frequent heavy precipitation, mountainous fringe zones, densely populated towns, and areas with concentrated cultivated land, while the northern regions are low-risk zones for rainstorm and flood disasters.

Keywords: Rainstorm · GIS · Disaster Zoning · Risk Index · Eastern End of the Hexi Corridor

1 Introduction

The spatial distribution characteristics of medium-to-heavy rainstorm frequencies in the eastern part of the Hexi Corridor show certain consistency with the spatial distribution of altitude and annual precipitation days. This is the result of the interaction among factors such as topography, climate, and climate systems. Meanwhile, the understanding of this distribution pattern is of great significance for local disaster prevention and control,

H.-S. Ma and Y.-W. Chan (Eds.): IC 2026, LNEE 1646, pp. 46–57, 2027.
https://doi.org/10.1007/978-981-92-0779-4_5

agricultural production, and other aspects. [1] Hu Ying [2] established a risk assessment model for rainstorm disasters in Yunnan Province by using the hourly precipitation data of 126 national meteorological stations in Yunnan Province from 2010 to 2019 and basic geographic information data. The model was constructed from three aspects: the hazard of disaster-causing factors of rainstorm disasters, the sensitivity of disaster-pregnant environment, and the vulnerability of disaster-bearing bodies. The entropy method was used to determine the weights, and the comprehensive value of rainstorm disaster risk was calculated through ArcGIS interpolation analysis and raster analysis. The natural break method was used to divide the whole province into five levels of rainstorm disaster risk. Ye Xiangyu et al. [3] constructed a risk assessment system for rainstorm and mountain flood disasters in the Sichuan-Tibet Transportation Corridor from three aspects: the hazard of disaster occurrence, the vulnerability of disaster-bearing bodies, and the social disaster reduction capacity. They used the spatial analysis functions of the GIS platform to carry out overlay analysis on multiple assessment indicators, and evaluated the risk of rainstorm and mountain flood disasters in the Sichuan-Tibet Transportation Corridor. The evaluation results were basically consistent with the actual disaster situations. This paper takes rainstorm disasters as the main research object, selects appropriate indicators, and constructs a meteorological disaster risk assessment model. It explores and studies the technical methods for rainstorm disaster risk zoning in the arid and semi-arid regions of the Hexi Corridor, determines the risk levels, and draws a risk grade zoning map for rainstorm disasters, so as to provide decision-making departments with a basis for formulating rainstorm disaster prevention measures and offer references for meteorological disaster risk zoning research in other regions [4].

2 Data Processing and Model Establishment

2.1 Normalization and Quantification of Rainstorm Disaster Risk Assessment Indicators

According to the conceptual framework of rainstorm disaster risk, 10 reference indicators are selected to evaluate the risk level of rainstorm disasters. Since the selected indicators have different units, in order to eliminate the dimensional and data level differences among the four evaluation factors of rainstorm disaster sensitivity, hazard, vulnerability and disaster prevention and mitigation capacity, first of all, it is necessary to carry out normalization processing on each indicator value [3]. The calculation formula is:

$$d_i = 0.5 + 0.5 \times \frac{x_i - x_{min}}{x_{max} - x_{min}} \tag{1}$$

In the formula (1): d_i refers to the normalized value of the *i-th* evaluation index, x_i refers to the value of the *i-th* evaluation index, x_{max} refers to the maximum value among all evaluation indices, and x_{min} refers to the minimum value among all evaluation indices (Fig. 1).

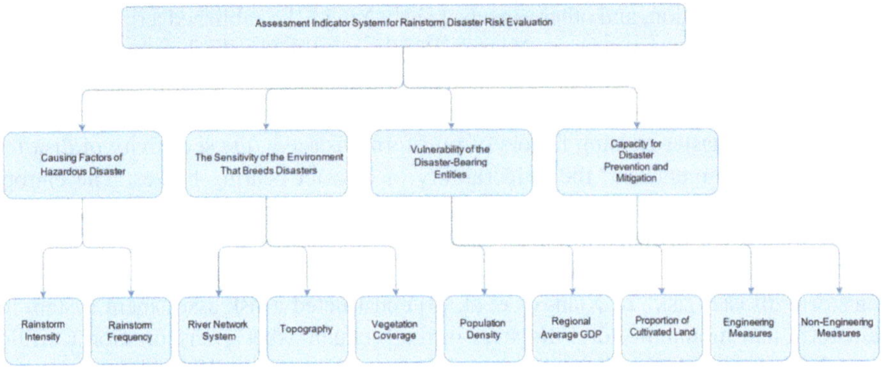

Fig. 1. Conceptual Framework of Rainstorm Disaster Risk

2.2 Establishment of Evaluation Model

Based on the conceptual framework of natural disaster risk, the Analytic Hierarchy Process (AHP) and weighted comprehensive evaluation method are adopted to construct a rainstorm disaster risk index model. Considering the different impacts of each indicator on the comprehensive evaluation factors of disasters, the model integrates the roles of specific indicators in disasters and then concentrates them into a quantitative index [4]. The calculation formula is:

$$f = \sum_{i=1}^{n} w_i \times d_i \tag{2}$$

In Eq. (2): f represents the value of the evaluation factor, d_i is the normalized value of the i-th indicator, w_i denotes the weight of the i-th indicator, n is the total number of evaluation indicators. The weight w_i reflects the relative importance of each evaluation indicator compared to the main factors contributing to disaster risk. Here, weights are determined through expert scoring methods [4].

2.3 Calculation of Evaluation Factor Weights

First, conduct a detailed investigation of the rainstorm disaster situation in the study area, and then carry out research and analysis on the distribution of disaster areas and influencing factors. On the basis of comprehensively considering the opinions of meteorological, water conservancy and geological experts in the study area, according to the natural disaster system theory and disaster risk assessment theory, select four aspects of hazard of disaster-causing factors, vulnerability of disaster-bearing bodies, sensitivity of disaster-pregnant environment and disaster prevention and mitigation capacity as the criterion layer, and select 10 evaluation factors corresponding to the criterion layer as the factor layer. Then, according to the contribution relationship between the evaluation factors, refer to Saaty's concept to construct the corresponding order judgment matrix. Taking the vulnerability of disaster-bearing bodies as an example, use the 1–9 scaling method to construct a third-order judgment matrix, and determine the relative importance of population density, GDP per unit area and cultivated land proportion through evaluation. The three are compared pairwise, and the final judgment value is an integer between 1–9, where 1 means that the two factors are equally important, the value increases in turn according to the degree of influence, and 9 means that one factor is extremely important compared with the other. Conversely, use the reciprocal to represent it. After comprehensively considering and determining the values of the judgment matrix, solve its maximum eigenvalue and its eigenvector, and carry out consistency test to obtain the weight values of each evaluation factor [5]. Substitute these values into the corresponding zoning superposition model. Use the Grid module of ArcGIS to carry out raster superposition operation and classification to obtain the regional distribution map of rainstorm disaster zoning grades.

2.4 Divide the Grade Boundaries

This paper uses the natural break method to divide the evaluation values into five grades, namely high-risk, sub-high-risk, medium-risk, sub-low-risk and low-risk areas (Table 1). The natural break classification method is a statistical formula used to determine the natural clustering of attribute values in order to reduce the differences within the same grade and increase the differences between different grades [4]. The calculation formula is:

$$ssdi - j = \sum_{k=i}^{j} (a[k] - meani - j)2(1 \leq i \leq j \leq n) \tag{3}$$

In Eq. (3): a is an array (with a length of k), and $mean_{i\text{-}j}$ represents the average value in each grade.

Table 1. Rainstorm risk grade standard

Grade	Area	Risk Index Range
Grade 1	High-risk Area	(1.00, 0.90]
Grade 2	Sub-high-risk Area	(0.90, 0.80]
Grade 3	Medium-risk Area	(0.80, 0.70]
Grade 4	Sub-low-risk Area	(0.70, 0.60]
Grade 5	Low-risk Area	(0.60, 0.50]

3 Zoning of Rainstorm Disasters in the Eastern End of the Hexi Corridor

3.1 Rainstorm Disaster Assessment Index System

The eastern end of the Hexi Corridor belongs to arid and semi-arid areas, with precipitation mainly concentrated in summer and autumn every year. During this period, heavy precipitation processes often cause rainstorm and flood disasters in this region.

According to the disaster risk theoretical model, when studying the rainstorm meteorological disasters in this region, four factors are comprehensively considered: the hazard of disaster-causing factors, the sensitivity of disaster-pregnant environment, the vulnerability of disaster-bearing bodies, and the capacity of disaster prevention and mitigation. Among them, the hazard of disaster-causing factors is mainly based on rainstorm intensity and frequency; the sensitivity analysis of disaster-pregnant environment mainly considers topographic factors (such as slope and height), river network density, and vegetation coverage; the vulnerability analysis of disaster-bearing bodies mainly focuses on population proportion, economic status, and cultivated land area; the capacity of disaster prevention and mitigation refers to a series of engineering and non-engineering measures taken to reduce losses caused by flood disasters.

3.2 Indicators and Analysis of Rainstorm Risk Assessment in the Eastern End of the Hexi Corridor

Analysis of Hazardousness of Disaster-Causing Factors.
Precipitation intensity and frequency are the main causes of rainstorms. Meanwhile, the spatial distribution of rainfall is also one of the factors affecting rainstorm disasters. Therefore, factors such as rainstorm intensity, duration, and spatial differences in rainfall are primarily considered when studying disaster-causing factors of rainstorms.

Hexi Corridor, located deep in the inland and far from the ocean, is affected by the terrain and altitude within its territory, resulting in long-term high temperatures, drought, scarce and uneven precipitation. Therefore, the criteria for rainstorm occurrence cannot be unified with those in the eastern regions [6]. According to the current meteorological business regulations of the Gansu Meteorological Bureau: in the Hexi Corridor region, a day with precipitation of 30 mm or more is defined as a rainstorm day [8], if two or more stations within a day have precipitation of $\geq$ 30 mm, it is identified as a regional rainstorm day.

Statistics were respectively conducted on the rainfall amounts of qualified rainstorm processes lasting 1 day, 2 days, 3 days,..., 10 days from 1961 to 2010 at each meteorological station in the eastern end of the Hexi Corridor. Then, the rainfall samples of the same duration from all meteorological stations in the eastern end of the Hexi Corridor were statistically analyzed (Table 2). In the table, R represents the precipitation amount with the unit of mm; "/" indicates that the sample is too small to have a specific value, thus being meaningless.

Table 2. Rainfall level of meteorological stations at the eastern end of Hexi Corridor [5]

Continuous days	Rainfall level				
	Grade 1 (60~70%)	Grade 2 (70~80%)	Grade 3 (80~90%)	Grade 4 (90~95%)	Grade 5(≥95%)
1	$26.6 \leq R \leq 34.6$	$34.6 \leq R \leq 42.7$	$42.7 \leq R \leq 53.0$	$53.0 \leq R \leq 63.6$	$R \geq 63.6$
2	$34.1 \leq R \leq 45.3$	$45.3 \leq R \leq 57.9$	$57.9 \leq R \leq 69.5$	$69.5 \leq R \leq 87.5$	$R \geq 87.5$
3	$43.3 \leq R \leq 57.5$	$57.5 \leq R \leq 71.9$	$71.9 \leq R \leq 82.7$	$82.7 \leq R \leq 99.0$	$R \geq 99.0$
4	/	/	/	/	/

Statistically analyze the occurrence frequency of different rainfall grades in each county and district at the eastern end of the Hexi Corridor, then sum the times of the same precipitation intensity across different precipitation cycles in each county/district to obtain the occurrence frequency of different grades of rainstorm intensity. According to the principle that higher rainstorm intensity grades have a greater impact on rainstorm disaster occurrence, the grade weights of rainstorm intensity are set as 1/15, 2/15, 3/15, 4/15, and 5/15 through the Analytic Hierarchy Process (AHP). Then, the weighted comprehensive evaluation method is adopted to calculate the sum of products of the precipitation intensity weights of different grades and the corresponding occurrence frequencies at 4 national meteorological stations, so as to obtain the hazard index of disaster-causing factors for rainstorm and flood disasters. In ArcGIS 10.0, the spatial interpolation method is used to draw the hazard grade distribution map of disaster-causing factors in the study area (Fig. 2), and the natural break method is applied to divide the hazard of disaster-causing factors into 5 grades [8].

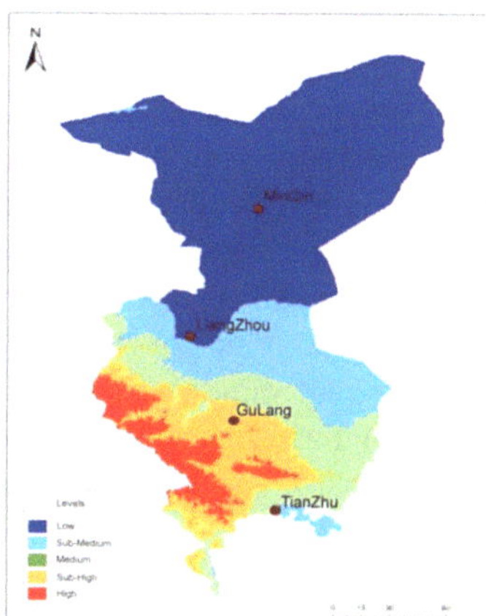

Fig. 2. Distribution of rainstorm hazard levels at the eastern end of Hexi Corridor

As shown in Fig. 2, the overall rainstorm risk zoning increases from north to south. The high-risk areas for rainstorms are mainly concentrated in the southern mountainous areas where heavy precipitation is frequent, as well as along mountainous regions, densely populated towns, and areas with concentrated cultivated land. The northern part is a low-risk area for rainstorms. This is because the southern mountainous areas have lower latitudes, and their water vapor conditions are superior to those in the north. Coupled with the uneven surface heating and dynamic uplifting effects caused by the large topography of the Qilian Mountains, water vapor condenses to form clouds and cause rainfall, resulting in more rainy days [6], making it a high-risk area for rainstorms. The Minqin area in the north is a desert region, so the number of precipitation days is the least, making it a low-risk area for rainstorms.

Analysis of Disaster-Pregnant Environment Sensitivity.
The terrain of the eastern end of the Hexi Corridor is high in the south and low in the north, with mountains, plains, deserts and gobi distributed in sequence from south to north. The mountainous areas in the eastern section of the Qilian Mountains have a high vegetation coverage rate, including Tianzhu, Gulang and the southern part of Liangzhou District. The central part of Liangzhou District, the area north of Gulang Gorge and the local part of the southeast of Tianzhu are dominated by semi-desert vegetation, with slightly higher vegetation coverage. In Minqin County, the northern part of Gulang County and some towns in the northeastern part of Liangzhou District, the climate is arid and the vegetation coverage rate is very low.

According to the causes of rainstorm disasters, the disaster-pregnant environment in the study area mainly considers three aspects: topographic factors (such as slope and height), river network distribution, and vegetation coverage.

(1) The Influence of River Network on Rainstorm Disasters

Generally, the higher the river network density and the closer the area is to water bodies such as rivers and reservoirs, the more prone it is to rainstorm disasters. The higher the river grade and the larger the water area, the greater the impact on rainstorm disasters. River network distribution mainly depends on river network density and river network buffer zones [5]. River network density is defined as the total length of rivers in a watershed per unit area, which can be calculated using the "Density" tool in the GIS spatial analysis module for the river network density within Wuwei City. According to the distance from rivers, reservoirs, and lakes, it is divided into two-level buffer zones, with a factor value of 0–1 selected. The principle is to assign large values to first-level rivers and large water areas, and small values to second-level rivers and small water areas for factor (factor assignment) in river network distribution. Finally, the influence index of river network distribution characteristics in the eastern end of the Hexi Corridor is calculated (Table 3).

Table 3. Criteria for division of river network buffer zone in the eastern end of Hexi Corridor [6]

Water Body Type		Buffer Zone Width/km	
		Level 1 Buffer Zone	Level 2 Buffer Zone
Level 1 River		8	12
Level 2 River		6	10
Water Area (10^4km^2)	[0.1,1)	0.5	1
	[1,10)	2	4

(2) The Influence of Terrain on Rainstorm Disasters

For the disaster-pregnant environment of rainstorm disasters, terrain is one of the most important determining factors. The terrain mainly selects the slope and altitude of the study area. The slope mainly considers the relative standard deviation of elevation within a certain distance around the grid unit, which can be used to reflect the degree of terrain change. The larger the value, it indicates that the terrain here is significantly different from the surrounding area. Therefore, rainstorm floods are not easy to occur in this area. On the contrary, rainstorm floods are prone to form. The altitude is also called the absolute height. The higher the absolute height and the smaller the relative standard deviation of elevation, the greater the sensitivity to rainstorm disasters. The grid layers of absolute height and relative standard deviation of elevation are superimposed and calculated, graded and assigned values, and finally the sensitivity influence index is obtained (see Table 4).

Table 4. Assignment table of terrain factors [5]

Terrain Elevation/m	Elevation Standard Deviation		
	Level 1(≤1)	Level 2(1~10)	Level 3(≥10)
Level 1(≤100)	0.7	0.8	0.9
Level 2(100~300)	0.6	0.7	0.8
Level 3(300~700)	0.5	0.6	0.7
Level 4(≥700)	0.4	0.5	0.6

(3) The Influence of Vegetation Coverage on Rainstorm and Flood Disasters.

Vegetation coverage mainly characterizes the water retention capacity of the study area. In general, the eastern end of the Hexi Corridor can be roughly divided into three distinct ecological zones from south to north: forest-steppe vegetation area, agricultural vegetation area, and desert vegetation area, with vegetation coverage decreasing gradually from south to north. The lower the vegetation coverage, the poorer the soil and water conservation capacity, and the greater the sensitivity to rainstorm disasters. Conversely, the higher the vegetation coverage, the stronger the soil and water conservation capacity, and the smaller the sensitivity to rainstorm disasters. Through the raster calculation function in ArcGIS software, the vegetation coverage distribution in the eastern end of the Hexi Corridor was calculated and divided into 5 levels by the natural break method.

Based on the influences of the above factors on the disaster-pregnant sensitivity of rainstorms in the eastern end of the Hexi Corridor, the weights of each index were calculated through the Analytic Hierarchy Process (AHP), overlay analysis was conducted in ArcGIS, and the results were graded by the natural break method, so as to obtain the sensitivity grade distribution map of the disaster-pregnant environment for rainstorms (Fig. 3).

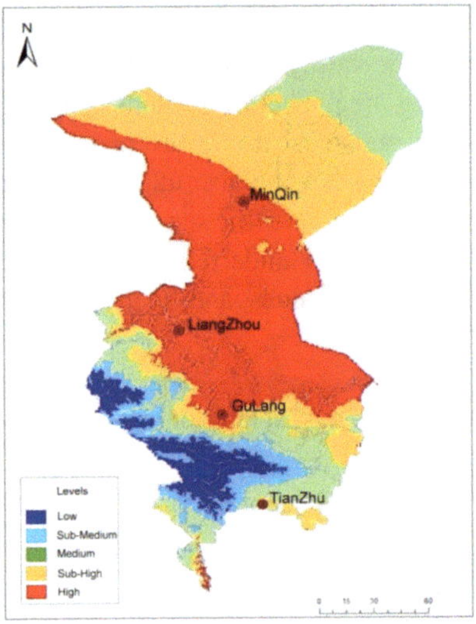

Fig. 3. Distribution of environmental sensitivity level of rainstorm hazard in the east end of Hexi Corridor

Vulnerability Analysis of Disaster-Bearing Bodies.
The vulnerability of disaster-bearing bodies for rainstorm and flood disasters in the eastern end of the Hexi Corridor mainly considers three factors: population, cultivated land, and economy. Based on the population density, GDP per unit land area, proportion of cultivated land in each region, as well as the actual situation obtained from local surveys. The weights of the above three indicators were calculated through the Analytic Hierarchy Process (AHP), overlay analysis was carried out in ArcGIS, and the results were graded by the natural break method to obtain the vulnerability grade distribution map of disaster-bearing bodies (Fig. 4).

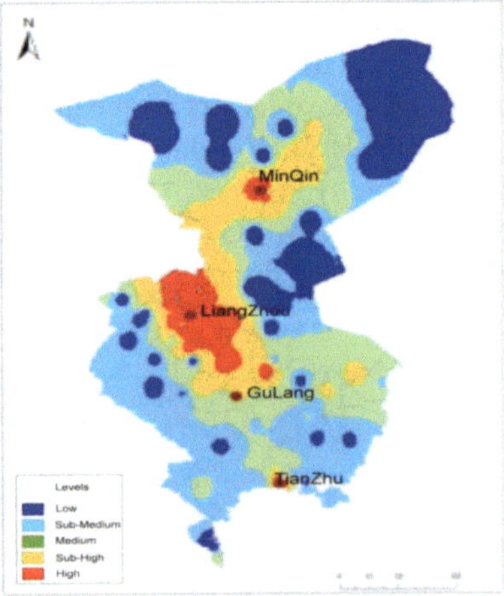

Fig. 4. Vulnerability distribution of storm disaster bearing bodies at the east end of Hexi Corridor

Analysis of Disaster Prevention and Mitigation Capability.
The analysis of disaster prevention and mitigation capability refers to the ability of disaster-affected areas to resist meteorological disasters, which is closely related to local emergency response capacity and economic conditions. A strong disaster prevention and mitigation capability means smaller potential losses and lower meteorological disaster risks. Economic conditions include local economic policies, per capita income, output value of the tertiary industry, etc., while emergency response capabilities mainly include the provision of water conservancy facilities, investment in emergency funds, and the quantity and quality of emergency personnel. Disaster prevention and mitigation capability belongs to subjective factors, and its specific quantification is relatively abstract without a unified evaluation standard, so it is generally not superimposed when drawing risk maps [4].

3.3 Risk Zoning and Analysis of Rainstorm Disasters in the Eastern End of Hexi Corridor

The risk assessment of rainstorm disasters mainly conducts comprehensive analysis from four aspects: the hazard of disaster-causing factors, the sensitivity of disaster-pregnant environment, the vulnerability of disaster-bearing bodies, and the capacity of disaster reduction and resistance. Considering the influence degree of each evaluation factor on the rainstorm disaster risk, the weights are assigned to each factor and then summarized. The following are the index weights of each application layer in the rainstorm disaster risk assessment model of the study area (Table 2). Since the hazard of disaster-causing factors, the sensitivity of disaster-pregnant environment, and the vulnerability of disaster-bearing bodies are positively correlated with the rainstorm disaster risk, while the disaster prevention and mitigation capacity is negatively correlated with the rainstorm disaster risk, it needs to be superimposed as a negative value if considered. Additionally, due to its strong subjectivity, only the former three are generally considered when drawing risk maps [4] (Table 5).

Table 5. Weight of rainstorm disaster risk assessment indicators

Target Layer	Criterion Layer	Weight	Evaluation Laye	Weight 1	Weight 2
Rainstorm Disaster Risk	Disaster-Causing Factors	0.46	Rainstorm Intensity	0.45	0.2070
			Rainstorm Frequency	0.55	0.2530
	Disaster-Pregnant Environment	0.18	River Network	0.29	0.0522
			Terrain	0.37	0.0666
			Vegetation Coverage	0.34	0.0612
	Disaster-Bearing Bodies	0.31	Population Density	0.31	0.0961

(continued)

Table 5. *(continued)*

Target Layer	Criterion Layer	Weight	Evaluation Laye	Weight 1	Weight 2
			GDP	0.18	0.0558
			Proportion of Cultivated Land	0.51	0.1581
	Disaster Prevention and Mitigation Capability	0.05	Engineering Measures	0.25	0.0125
			Population Quality	0.34	0.0170
			Non-Engineering Measures	0.41	0.0205

According to the established rainstorm disaster risk assessment model in the eastern end of the Hexi Corridor, spatial analysis and raster calculation functions in ArcGIS software are used to carry out superposition analysis on the three factors of hazard risk, disaster-inducing environment sensitivity and disaster-bearing body vulnerability according to the corresponding weights, and the natural breakpoint method is used to divide the disaster into five-level zoning (as shown in Fig. 5).

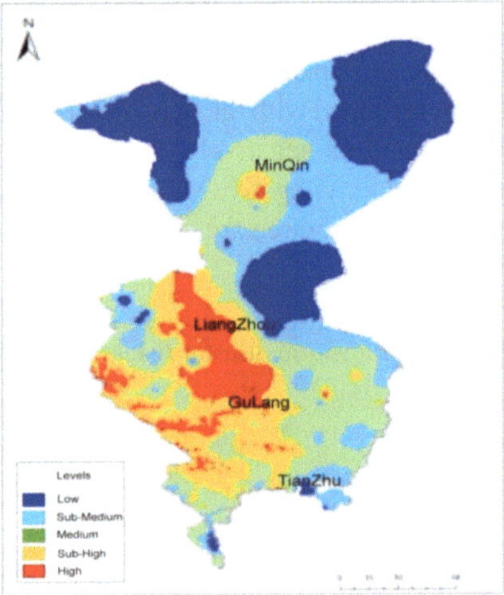

Fig. 5. Rainstorm risk grade zoning map of the eastern end of Hexi Corridor

4 Conclusion

(1) The high-risk areas of the disaster-causing factors of rainstorm disasters in the eastern end of the Hexi Corridor are mainly concentrated in the southern regions with relatively high precipitation, most parts of Minqin and Liangzhou Districts, where the risk of triggering rainstorm disasters is low.

(2) The high-sensitivity areas of the disaster-inducing environment for rainstorm disasters in the eastern end of the Hexi Corridor are mainly distributed in the central plain areas of the eastern end of the Hexi Corridor, followed by the Minqin desert area with low vegetation coverage; the southwestern part of Tianzhu has high vegetation coverage, and it belongs to high-altitude areas, so the sensitivity of the disaster-inducing environment is relatively low.
(3) Liangzhou District has a dense population and relatively good social and economic development, so its vulnerability is relatively high. The low-vulnerability areas are mainly distributed in the eastern and western parts of Minqin County, the eastern part of Gulang County, and the eastern part of Tianzhu County.
(4) The high-risk areas for rainstorm disasters in the eastern Hexi Corridor are primarily concentrated in the main urban areas of each county/district, the southern part of Liangzhou District, and most parts of Gulang County. Tianzhu County lies in the medium-high risk zone, while other regions generally face lower risks, with Minqin County having the lowest risk. Overall, the high-risk zones for rainstorms include the southern mountainous areas prone to heavy precipitation, foothill regions, densely populated towns, and areas with concentrated farmland, whereas the northern regions are classified as low-risk zones for rainstorm-induced floods.

Fund. 2025 Gansu Provincial University Teachers' Innovation Fund Project (No. 2025A-305). This research was supported by the Science Popularization Capacity Building Plan of the Gansu Provincial Technology Innovation Guidance Program (Grant No. 26CXKA023).

References

1. Haibei, C., Xiaoling, Y., Huihua, Z., et al.: Analysis of a sudden heavy rain weather event in the Eastern Part of the Hexi Corridor. South-Central Agric. Sci. Technol. **44**(05), 174–178 (2023)
2. Ying, H., Xian, Y., Jianqiao, C., Hua, Y., Zhifang, D.: 1km Refined rainstorm disaster risk assessment in Yunnan Province Based on GIS. Meteorol. Sci. Technol. **50**(5), 742–750 (2022)
3. Xiangyu, Y., Yuhan, G., Liaofeng, L., Jiayu, T., Zhonggen, W., Zhen, H.: Risk assessment of rainstorm and mountain flood disasters in Sichuan-Tibet Transportation Corridor. **32**(4), 48−60 (2023)
4. Yunfeng, D., et al.: Risk Assessment and Zoning of Hail Disasters in the Eastern Qilian Mountains Based on GIS. J. Yangtze River Sci. Res. Inst. **05** (2016)
5. Ya, X.: Research on Risk Assessment Technology of Main Meteorological Disasters Based on GIS Technology. Guangxi Normal University, Guilin (2014)
6. Li, Q., Yonglong, Y., Yulin, Z., Jie, T., Juju, L.: Analysis of Spatiotemporal Distribution and Influencing Factors of Extreme Precipitation in Eastern Hexi Corridor. Arid Land Geogr. **38**(02), 207–214 (2015)
7. Collection of Technical Regulations for Surface Observation Business, Gansu Meteorological Bureau, pp. 51–52 (2009)
8. Hongying, L., Xiaoyu, Z., Ning, C., et al.: Comparative analysis of two drought indices in the Hazard Risk of Drought-Inducing Factors. J. Catastrophology **27**(2), 58–61 (2012)

AI/AR Smart Glasses in Nursing: Frontiers, Risks, and Governance

Chao-Ting Chu[1], Hsiung-Mao Lin[2], Yueh-Tao Chiang[3,4], and Wen-Yen Lin[4,5(✉)]

[1] 1Internet of Things Laboratory, Chunghwa Telecom Laboratories, Taoyuan, Taiwan
[2] 2Department of Computer Science and Information Engineering, National Taichung University of Science and Technology, Taichung, Taiwan
[3] 3Department of Nursing, College of Medicine, Chang Gung University, Taoyuan, Taiwan
[4] Division of Pediatric Endocrinology and Genetics Department of Pediatrics, Chang-Gung Memorial Hospital, Taoyuan, Taiwan
qqnice@gmail.com
[5] Department of Information Management, National Taichung University of Science and Technology, Taichung, Taiwan

Abstract. We examine AI/AR smart-glasses integration in nursing from a governance lens. This work is a narrative review and conceptual framework that searched IEEE, ACM Digital Library, Scopus, and Web of Science for publications from 2013–2025. Search terms were smart glasses, augmented reality, mixed reality, nursing, nurse, nursing education, education, learning, ethics, privacy, and governance. Inclusion criteria were studies with identifiable methods addressing educational outcomes, usability, cognitive load, or data governance. Exclusion criteria were papers without methods or results and duplicates. The evidence indicates benefits for bedside communication, visualization, decision support, and education, alongside risks involving privacy and security, human factors, and workflow fit. We outline a layered socio-technical governance approach linking design, policy, implementation, and technical safeguards. Operationalization emphasizes on-device processing, access control, encryption, and auditability to support safe, ethical, and effective use in nursing settings. These findings provide a synthesis and a roadmap. No definitive evidence grading is claimed.

Keywords: Smart glasses · Augmented reality · Nursing practice · Healthcare applications · Privacy · Data governance

1 Introduction

Health systems are undergoing digital transformation amid population aging and rising clinical complexity. AI/AR smart glasses combine head-up displays and hands-free interaction to overlay step guidance, medication verification, and physiologic parameters, while first-person streaming supports remote collaboration and preceptorship, narrowing the gap between classroom and ward. Yet prior reviews have treated applications and barriers separately, leaving the field without an integrated model to guide implementation. Accordingly, this paper moves beyond description and presents a socio-technical governance framework aligned with clinical workflows and human factors [1–4].

H.-S. Ma and Y.-W. Chan (Eds.): IC 2026, LNEE 1646, pp. 58–63, 2027.
https://doi.org/10.1007/978-981-92-0779-4_6

2 Technology Landscape for Healthcare Smart Glasses

2.1 Modalities and Trade-Offs

Early devices were constrained by battery life, imaging quality, heat, and privacy. With AI and sensing, devices now provide context-aware recognition and personalized assistance. Two paths dominate:

Assistive AR (monocular): Optimized for prompts and endurance.

Mixed reality (binocular): Optimized for spatial understanding and 3D interaction.

Clinical selection balances human factor load, field of view, endurance, and the need for spatial perception [3, 5, 6]. In typical nursing scenarios, monocular assistive AR is favored for endurance and unobstructed vision, whereas binocular mixed reality is reserved for tasks requiring depth perception and 3D interaction (Table 1).

Table 1. Platform Snapshots.

Name	Type	Key Tech	Inputs	Status/Use
Meta Ray-Ban Display [7]	AI smart glasses with in-lens display	HUD display, Meta AI	Voice, Meta Neural Band (EMG) gestures	Consumer. Announced 2025–09. Navigation, Translation, Messaging, Camera [7]
Android XR [8]	OpenXR-compliant platform	Multi-device, Gemini AI / Google apps integration	Device-dependent	VR anatomy, Digital therapeutics (pain/mental health) [9, 10]
MediThinQ MetaSCOPE [11]	3D surgical loupe	HUD, near-zero-latency wireless	Viewing via HMD	OR visualization, Ergonomic considerations [11]
Surgical ARVIS® [12]	*AR joint navigation*	Head-worn, IR tracking	Visual overlays	FDA 510(k) cleared, implant guidance [13]

Trade-offs: General-purpose platforms favor extensibility and fast iteration but typically incur greater clinical validation and regulatory overhead. Medical-grade systems foreground sterility, reliability, and compliance, yet demand higher total cost of ownership and steeper training thresholds. This split calls for risk-tiered validation, data-protection baselines, and role-aligned training to balance speed and safety.

3 Applications in Nursing Practice and Education

3.1 Clinical Nursing Practice

Communication in emergencies and isolation care: Clearer, quicker interaction between staff inside and outside the isolation room. Junior nurses receive real-time guidance, reducing delays and improving safety [14].

Procedures and patient safety: Hands-free display benefits aseptic, two-handed tasks. Ultrasound visualization in the line of sight raises first-pass success and shortens procedure time, reducing complications [15–17]. For negative pressure wound therapy, AR guidance with 3D cues supports risk reduction and flow optimization [18].

Decision-making and efficiency: Head-up display reduces eye/head switching and stabilizes performance in cannulation and related tasks [16, 17]. First-person streaming enables faster remote input and cross-site decisions [16, 19]. In anesthesia care, live vital signs in the field of view strengthen situational awareness [2, 16]. Evidence is early and context-dependent. Cross-setting replication is required before scale.

3.2 Nursing Education and Training

Accelerating skill acquisition: Supports self-practice. More repetitions correlate with higher satisfaction and shorter completion times [20].

Immersive, high engagement learning: Compared with traditional methods, smart-glasses-enabled simulation improves motivation, hands-on translation of theory, and retention [21, 22].

Reducing anxiety and building confidence: Extended-reality simulation provides a safe, repeatable environment for decision practice, while gesture- and procedure-specific rehearsal deepens immersion and a sense of control, which lowers stress and improves confidence. Yet because these tools place information directly in the nurse's line of sight, a robust socio-technical governance framework is required to manage the associated risks [23, 24].

4 Risks and Challenges

Technical/hardware: Weight, heat, small displays, and latency raise fatigue and timeliness risks. Gloves can cause touch misfires. Noise degrades speech recognition [1, 3, 20, 25].

Human/social: Reduced eye contact can hurt trust. Cameras raise surveillance concerns. Learning curve, privacy worry, and overload add strain. Liability boundaries blur with opaque algorithms [25, 26].

Privacy and data security: Recording without explicit consent risks privacy violations. Interception and device loss can expose sensitive data [4, 25, 27].

Organization/workflow: Poorly planned pilots can add steps, load, and new errors. TCO includes upkeep, updates, access control, and IT support. Connectivity and continuous training are prerequisites. Start with low-risk educational or adjunct scenarios to build operational evidence before broader rollout [1, 2, 25].

5 Mitigation Strategies and Governance

User-centered design: Engage frontline nurses early. Tune weight, UI flow, and inputs to work rhythms. Use context-aware AI to surface only crucial information. Strengthen multimodal input (noise-robust speech, low-amplitude/high-precision gestures) [3, 6, 28]. Evaluate cognitive load with validated measures and compare monocular versus binocular conditions in scenario-based tests.

Policy and ethics: Define permitted and prohibited scenarios and require informed consent. Enforce end-to-end encryption and strict access control with multi-factor authentication (MFA). Where feasible, use federated learning. Safeguard dignity, trust, and equity. Offer granular and revocable consent with patient-facing privacy dashboards [4, 6, 25].

Implementation and education: Comprehensive curricula should cover device operation, privacy/ethics, and patient communication. Transparent explanations of purpose and data handling increase acceptance and reduce anxiety [1, 29, 30]. Include critical technology literacy to prevent over-reliance, covering uncertainty interpretation, bias awareness, and explicit model-override protocols.

Technical safeguards and governance stack: Prioritize edge processing, strict data minimization, and explicit retention limits. De-identify uploads and encrypt transport. Enforce versioning and data lineage. Adopt strong identity and access management with device–cloud trust chains. Ensure integrity with cryptographic signatures and hashes. Use federated learning with differential privacy. Harden against adversarial inputs. Protect models and keys in trusted execution environments with sound key management [4, 6, 31]. Run latency-critical protected health information(PHI) on device. Buffer, de-identify, and encrypt non-urgent workloads for on-premises or trusted-cloud processing under strict access controls. For decision support, require explainability with case-level rationales and calibrated uncertainty. Limit opaque models to low-risk administrative use only.

6 Conclusion

AI/AR smart glasses are moving from display-centric wearables to context-aware aids in nursing. This review maps the device landscape and trade-offs, synthesizes early evidence for communication, procedures, and education, and identifies recurring risks in human/social interaction, privacy and security, and workflow burden. Based on these observations, we sketch a provisional socio-technical governance outline. Generality is not claimed. Adaptation to local contexts and empirical evaluation for effects on patient outcomes and nurse–patient relationships are required. Near-term priorities include risk-tiered offloading, granular consent design, and critical technology literacy to operationalize the framework in pilot deployments. Subject to careful governance and ongoing review, integration should prioritize patient safety, process efficiency, and educational quality.

References

1. Wüller, H., et al.: A scoping review of augmented reality in nursing. BMC Nurs. **18**(1), 19 (2019)
2. Elrose, F., et al.: The use of head-worn displays for vital sign monitoring in critical and acute care: systematic review. JMIR Mhealth Uhealth **9**(5), e27165 (2021)
3. Baashar, Y., et al.: Towards wearable augmented reality in healthcare: a comparative survey and analysis of head-mounted displays. Int. J. Environ. Res. Public Health **20**(5) (2023)
4. Elendu, C., et al.: Legal implications for clinicians in cybersecurity incidents: a review. Medicine (Baltimore) **103**(39), e39887 (2024)
5. Kim, D., Choi, Y.: Applications of smart glasses in applied sciences: a systematic review. Appl. Sci. **11**(11), 4956 (2021)
6. Wang, B., et al.: A systematic literature review on integrating AI-powered smart glasses into digital health management for proactive healthcare solutions. npj Digital Med. **8** (2025)
7. Meta: Meta Ray-Ban Display: AI Glasses With an EMG Wristband. (2025). [cited 20 Sep 2025]. Available from: https://about.fb.com/news/2025/09/meta-ray-ban-display-ai-glasses-emg-wristband/
8. Android Developers. Android XR. [cited 25 Aug 2025]. Available from: https://developer.android.com/xr
9. Xrhealth. Xrhealth. XRHealth. Available from: https://www.xr.health/us/products/
10. Virtual, M.: Human Anatomy VR: The world's most advanced VR anatomy platform. Virtual Medicine [Web Page]. Available from: https://www.medicinevirtual.com/
11. MediThinQ Co, L., MetaSCOPE Product Catalog v4.3. MediThinQ Co., Ltd.
12. Enovis. ARVIS®. [cited 31 Aug 2025]. Available from: https://enovis.com/products/surgical/arvisr
13. U. S. Food&Drug Administration. ARVIS® Shoulder: 510(k) Premarket Notification (K240062). [cited 30 Sep 2025] (2024). Available from: https://www.accessdata.fda.gov/scripts/cdrh/cfdocs/cfpmn/pmn.cfm?ID=K240062
14. Park, S., Lee, H.: The effect of communication in emergency department isolation rooms using smart glasses: a mixed-methods study. J. Clin. Nurs. (2025)
15. Tracey, M.-A., Miller, A., Clifford, K.: An Educational Module Utilizing Smart Glass Technology as an Adjunct During Anesthesia and Procedural Tasks to Decrease Medical and Human Errors in the Perioperative Period (2023)
16. Wang, Y., et al.: The effect of smart glasses combined with ultrasound on radial arterial catheterization: a randomized controlled trial. BMC Anesthesiol **24**(1) (2024)
17. Romare, C., Skär, L.: Smart glasses for caring situations in complex care environments: scoping review. JMIR Mhealth Uhealth **8**(4), e16055 (2020)
18. Wei, W., et al.: A scoping review of the applications of augmented reality in nursing. BMC Nurs. **24**(1) (2025)
19. Apiratwarakul, K., Cheung, L.W., Ienghong, K.: Impact of smart glasses on patient care time in emergency medical services ambulance. Prehosp. Disaster Med. **38**(6), 735–739 (2023)
20. Kim, S.K., et al.: Adaptation of extended reality smart glasses for core nursing skill training among undergraduate nursing students: usability and feasibility study. J. Med. Internet Res. **23**(3), e24313 (2021)

21. Alkhaqani, A., Rahi Al-Murshedi, A., Kadhim, J.: Smart Glasses Technology in Nursing Education: A New Tool in Simulation. AI Nurs. Res. Pract. pp. 16–23 (2025)
22. Johnston, M., et al.: The feasibility and usability of mixed reality teaching in a hospital setting based on self-reported perceptions of medical students. BMC Med. Educ. **24**(1) (2024)
23. Pérez De Los Cobos Cintas, C., et al.: Effect of an extended reality simulation intervention on midwifery students' anxiety: systematic review. JMIR Nurs. **8**, e68984 (2025)
24. Romare, C., Skär, L.: The use of smart glasses in nursing education: a scoping review. Nurse Educ. Pract. **73**, 103824 (2023)
25. Zuidhof, N., et al.: Perspectives on the acceptance and social implications of smart glasses: a qualitative focus group study in healthcare. Int. J. Hum.-Comput. Interact. **40**(2), 149–159 (2024)
26. Price, W.N., II, Gerke, S., Cohen, I.G.: Potential liability for physicians using artificial intelligence. JAMA **322**(18), 1765–1766 (2019)
27. Vijayan, V., et al.: Review of wearable devices and data collection considerations for connected health. Sensors **21**(16), 5589 (2021)
28. Zhang, Z., et al.: A smart glass telemedicine application for prehospital communication: user-centered design study. J. Med. Internet Res. **26**, e53157 (2024)
29. Odenheimer, S., et al.: Patient acceptance of remote scribing powered by google glass in outpatient dermatology: cross-sectional study. J. Med. Internet Res. **20**(6), e10762 (2018)
30. Press, V., et al.: Patient perceptions of wearable face-mounted computing technology and the effect on the DoctorPatient relationship. Appl. Clin. Inform. **07**(04), 946–953 (2016)
31. Johns, M., et al.: Data provenance in biomedical research: scoping review. J. Med. Internet Res. **25**, e42289 (2023)

An MLOps Framework for Real-Time BI-RADS Classification

Wan Mohd Amir Fazamin Wan Hamzah[1,2](✉), Nur Farah Afifah Ahmad Sukri[1], Mohd Kamir Yusof[1], Mumtazimah Mohamad[1,2], Ismahafezi Ismail[1], and Harmy Mohamed Yusoff[3]

[1] Faculty of Informatics and Computing, Universiti Sultan Zainal Abidin, Terengganu, Malaysia
amirfazamin@unisza.edu.my

[2] Artificial Intelligence Research Centre for Islam and Sustainability (AIRIS), Universiti Sultan Zainal Abidin, Terengganu, Malaysia

[3] Faculty of Medicine, Universiti Sultan Zainal Abidin, Terengganu, Malaysia

Abstract. This paper presents a practical framework that delivers real-time BI-RADS classification and maintains the model's reliability after deployment. A TensorFlow classifier runs behind a small web service that accepts up to eight mammogram images simultaneously and returns the predicted BI-RADS label along with its corresponding probability. Each request is tagged with a timestamp, a request ID, and the exact model version, allowing results to be traced end-to-end. Operational health is tracked through a set of simple signals. The service reports latency percentiles for a five-minute window and is evaluated against a clear target, such as a p95 of under 300 ms. It also monitors the mix of predicted classes over time and estimates distribution shifts using two standard scores: the Population Stability Index (PSI) and the Jensen–Shannon divergence (JSD). Clinicians can accept or reject predictions; accepted cases form a growing dataset. When enough verified examples have accumulated or the drift crosses a limit, the system initiates a background training job with guardrails to prevent overlap and stale runs. New models are recorded in a registry and served automatically. Over the last ~ 20 training cycles, validation accuracy held in the mid-90s, reaching a high of roughly 94.6%. In tests on a local deployment, the framework kept p95 latency well below the target. It refreshed the model using only newly accepted cases, reducing manual effort and shortening the time from data collection to improved predictions. The result is a fast, observable, and self-improving pipeline that is well-suited to day-to-day breast imaging workflows.

Keywords: MLOps · BI-RADS · Real-Time Inference · Drift Monitoring · Auto-Training

1 Introduction

Breast imaging services rely on transparent, consistent reporting so that findings lead to appropriate follow-up. The Breast Imaging Reporting and Data System (BI-RADS) provides a shared language; however, mapping images to categories at the workstation

H.-S. Ma and Y.-W. Chan (Eds.): IC 2026, LNEE 1646, pp. 64–74, 2027.
https://doi.org/10.1007/978-981-92-0779-4_7

remains challenging [1]. Models that perform well offline may slow down under load, drift as populations and protocols change, or become hard to trust without transparent operational signals.

This study addresses three practical problems. First, response time must remain low during busy sessions; a slow assistant goes unused. Second, model behaviour changes over time with scanners, workflows, and case mix; performance can degrade quietly without continuous checks. Third, learning from daily practice should be safe and traceable: when a clinician confirms a prediction, the case should strengthen the model only under clear safeguards, with a record of what ran, when, and why.

The framework integrates a TensorFlow classifier with a web-based system for online inference, and a monitoring/automation layer to meet these needs. The system records per-request timing, tracks uptime, compares recent and baseline class distributions using the Population Stability Index (PSI) and Jensen–Shannon divergence (JSD) [2], and surfaces the results on an Operations Console. When accepted cases reach a threshold or drift exceeds limits, the system launches a guarded retraining job and logs its status end-to-end.

2 Related Works

Clinical mammography has long relied on BI-RADS to standardise descriptions and actions. The fifth edition clarified categories and descriptors to reduce variation across readers and sites, which is crucial when decisions such as short-interval follow-up or biopsy depend on consistent language [3]. Structured reporting improves auditability and quality improvement, but manual assignment is time-consuming and sensitive to workload.

Deep learning shows strong promise for BI-RADS-aligned tasks. Convolutional models can classify screening mammograms, flag suspicious regions, and support triage; several studies report competitive or complementary performance to readers when used as decision support [4–7]. Category-prediction studies indicate that networks trained with curated labels and robust augmentation can mirror BI-RADS patterns, especially when paired with view and prior-exam context.

Moving from a trained model to a dependable system introduces MLOps concerns. Early accounts of "technical debt" in ML emphasised that data pipelines, configuration, and feedback loops can fail silently even when model code is tidy [8]. Platforms such as TensorFlow Extended and MLflow add structure for pipelines, metadata, versioning, and reproducibility [9, 10], which is essential for audit trails and safe rollouts in clinical settings.

Change over time is central. Surveys show that relationships between inputs and labels shift with scanners, protocols, demographics, and policy; left unmonitored, calibration and accuracy erode [11]. Practical monitoring often compares recent and operational indices, such as PSI and baseline distributions, using information-theoretic distances, like JSD [12]. PSI summarises how recent class shares deviate from a baseline across categories. Small values indicate stable populations, values around 0.1–0.25 suggest meaningful change, and larger values signal a substantial shift that may warrant model review or retraining. While JSD quantifies the difference between two class distributions by comparing each to its average, it is symmetric, bounded (with 0 indicating no difference; larger values indicating a greater shift), and remains stable even when some classes are rare. Deployment reports pair such statistics with guardrails alerts, cooldowns, validated scripts, and locks to balance responsiveness with safety.

3 Methods

The framework follows a closed loop, as shown in Fig. 1. The Operations Console sends images to Online Inference for BI-RADS predictions. Verification accepts or rejects each case. Accepted cases are accumulated in an Accepted Dataset, and all prediction records are stored in a Predictions database (DB). Monitoring exposes four signals (latency, uptime, accepted-case count, and drift). A Drift & Threshold Decision node authorises Auto-Train when limits are crossed or enough verified data have been gathered. The resulting model is written in a model registry and selected by the model serving system for deployment back to online inference. The Console can also issue a manual trigger.

3.1 Online Inference

A system endpoint wraps the TensorFlow classifier. A request may contain up to eight mammograms. Images are resized to the expected input shape and passed to the predictor. The response includes probabilities for all classes, the top BI-RADS label, a UTC timestamp, the model version, route, and a unique request ID. Per-image latency is measured in milliseconds and logged for aggregate percentiles and histograms. All mammography images entering the online inference are de-identified before processing. Direct patient identifiers such as names, identification numbers, and contact information are removed at the clinical system level, and only de-linked imaging data and associated BI-RADS labels are stored in the training and monitoring environment.

3.2 Verification and Accepted-Data Curation

Clinicians mark each case as Accept or Reject. Accepted cases are saved with compressed image bytes and label; when raw bytes are available, the image is also written to a class-specific folder under the BIRADS dataset for future training. A persistent accept count increases with each accepted case and acts as one gate for retraining.

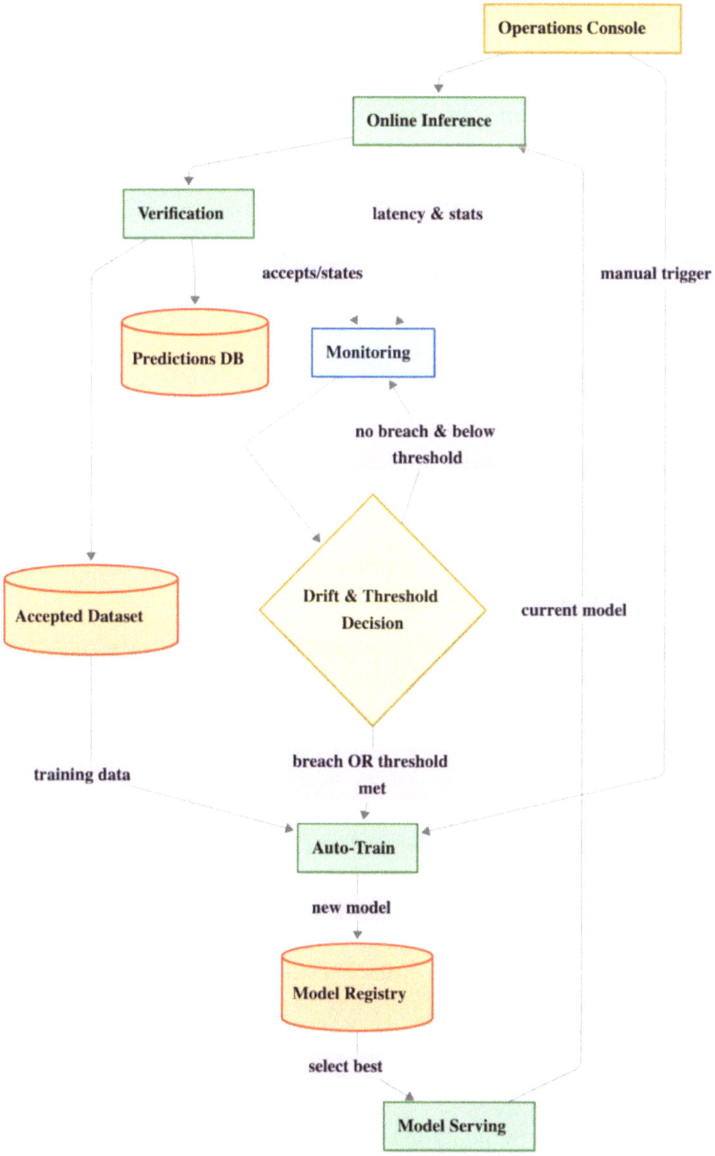

Fig. 1. An MLOps Framework for Real-Time BI-RADS Classification

3.3 Monitoring

Four small HTTP endpoints expose operational health:

- Latency returns recent records, UTC, route, model name, worker host, request ID, elapsed milliseconds, server-side percentiles (p50, p90, p95, p99), and a bucketed histogram. In this study, percentiles summarise the full distribution of request latency rather than relying on a single average. The p50 (the median) is the latency at or below which half of requests complete; p90 is the latency met by 90% of requests; p95 captures the upper tail where occasional slowdowns appear; and p99 probes the extreme tail most sensitive to rare contention or cold starts.

- Uptime publishes the app start time and host uptime to distinguish between clean restarts and slowdowns.
- Retraining state reports the current accept count, the configured threshold, and the most recent training job's status (start/finish times, return code, and lock presence).
- Data drift compares the distribution of predicted BI-RADS classes in a recent window with a baseline window and returns PSI, JSD, and the exact class share vectors and timestamps used

3.4 Drift and Threshold Decision

Two independent triggers can authorise a new training job:

- Accepted sample trigger. Start training when the accept count reaches the threshold (default 50).
- Drift trigger. A watchdog (defaulting to every 5 min) recomputes PSI and JSD for recent versus baseline windows. A retrain is allowed if either score exceeds its limit (defaults: PSI > 0.20 or JSD > 0.10).

Both triggers observe guardrails: a cooldown interval between runs, a file lock to prevent overlap, and a small healer that clears stale state if a prior process died. The Operations Console may also issue a manual trigger when needed.

3.5 Auto-Train

When an allowed trigger fires, the server launches a background training subprocess using the configured script. The worker records the last start time, last finish time, return code, and PID in a status file and mirrors the minimal state into a lock file for liveness checks. The accept count resets at job start to prevent repeated triggers during the same run. If a job exceeds its time-to-live or the PID disappears, the healer marks the run as failed and clears the lock, allowing future jobs to proceed.

3.6 Model Serving

A successful run writes a new model and a row in the model versions table with validation accuracy and a timestamp. At application start and whenever a new model is registered. Model Serving selects the best available version: first from the registry (highest accuracy, most recent), then from the newest artefact on disk, with a final fallback to a packaged default. A small in-process cache keeps loads fast and reduces cold starts. The current model identity is surfaced to the Console and attached to every prediction, enabling full traceability.

4 Results

4.1 Prediction

The prediction results in Fig. 2 show where images are scored in real-time and then immediately reviewed. The "Prediction" line reports the single BI-RADS category the model considers most likely after analysing the image. The classes are birads1–birads5, which broadly map to the clinical scale: birads1 (negative), birads2 (benign finding), birads3 (probably benign; short-interval follow-up often considered), birads4 (suspicious; biopsy should be considered), and birads5 (highly suggestive of malignancy). The result is birads2, which indicates the model saw patterns typical of a benign finding and that routine screening would generally be appropriate. The choice is made by comparing the model's probability for each class and selecting the one with the highest value.

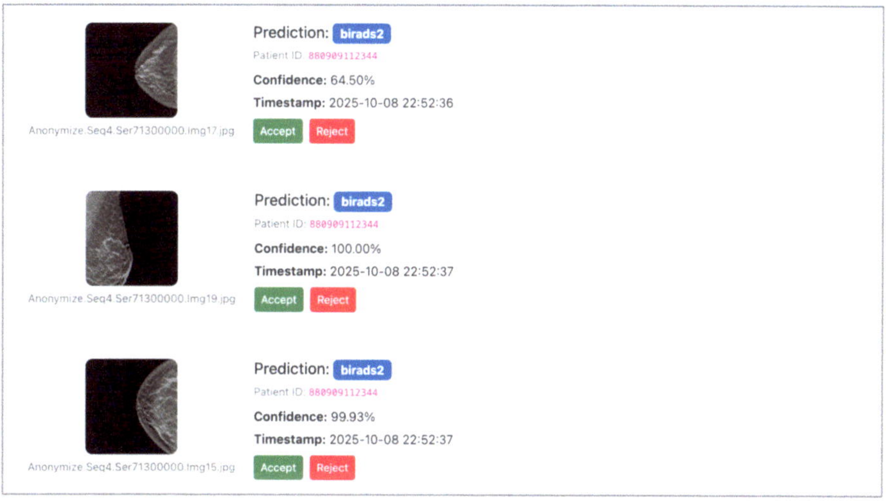

Fig. 2. Prediction Score Results

The confidence value is the model's estimated probability for the shown label. A higher number indicates that the model identified strong visual cues for that category; a lower number indicates that the evidence was mixed and the decision was more uncertain. Confidence aids judgment, not a promise; two cases with the same label but different levels of confidence may warrant different follow-up.

Selecting Accept records that the displayed label matches the clinician's assessment for this image. The system stores a compressed copy of the image in the "accepted" dataset and increments an internal counter of verified cases. That counter is used to determine when sufficient new, curated examples exist to justify initiating a new training run, so each Accept confirms the current case and contributes to an improved model later.

Selecting Reject records for which the model's label should not be used for learning from this case. Common reasons include motion blur, a mislabeled or duplicate image, a view that does not align with the model's expectations, or an apparent discrepancy with the clinical reading. If a case was accepted earlier and later deemed unsuitable, Reject removes it from the accepted dataset and lowers the verified-case counter, ensuring the training set remains clean.

4.2 Monitoring and Analysis

The MLOps monitoring and analysis in Fig. 3 tell a clear story. If response time drifts past the target, the alert switch can notify the team immediately. If the class mix starts to shift, the drift scores will flag it before accuracy drops in daily use. Meanwhile, the steady growth of accepted cases moves the system toward the next training cycle, and a single click can start it sooner when needed. By combining fast feedback (latency), steady background checks (uptime), controlled data growth (accepted samples), and simple drift signals (PSI and JSD), the page shows when the service is healthy and when it needs attention without digging through logs or raw tables.

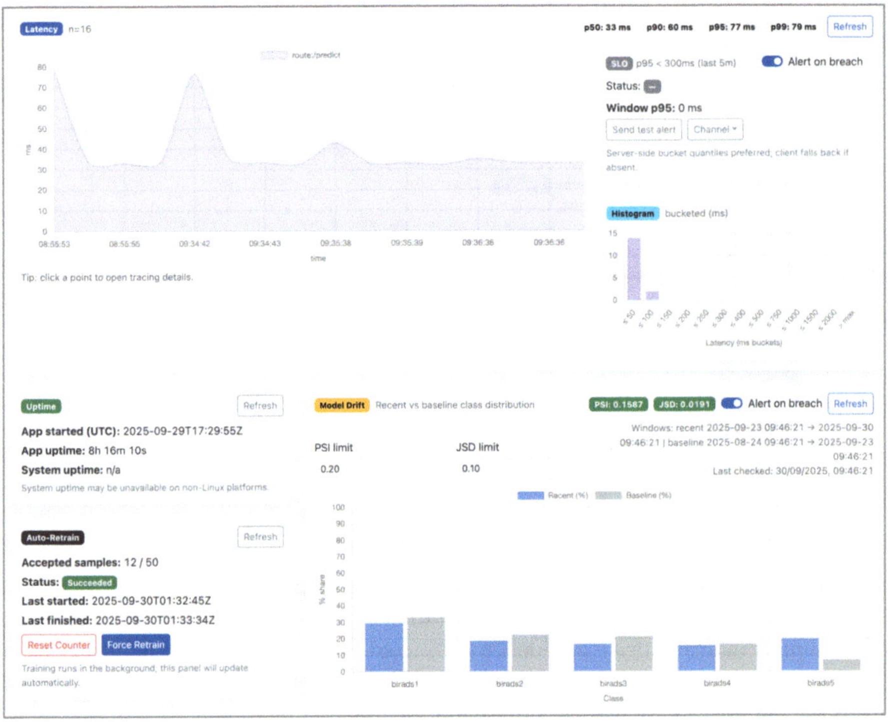

Fig. 3. MLOps monitoring and analysis

- The purple line in the Latency graph indicates the time it took the system to respond to each recent prediction request. The small numbers in the header give a quick read on typical and slow cases: half of the requests finished in about 33 ms (p50), nine out of ten in 60 ms (p90), and even the slower tail (p95 and p99) stayed under 80 ms. The counter n = 16 says there are sixteen points in this view.

- Service Level Objective (SLO) checks the service against a clear promise: "p95 under 300 ms in the last five minutes." The window's p95 is 58 ms, so the badge displays 'Good'. If that number rises above the target, switching on Alert on breach will send a message to the chosen channel; the Send test alert button confirms the path works before relying on it. Besides the badge, a compact histogram groups requests by speed. Most bars sit in the first buckets, showing the service spends nearly all its time under 100 ms.
- The Uptime panel provides context for any blips seen elsewhere. It shows when the app started, 2025-09-29T17:29:55Z, and how long it has been running, 8 h, 6 min, and 57 s, at the moment of capture. If there had been a restart a few minutes earlier, a brief period of slower responses might be expected while caches warm up.
- The Auto-Retrain block tracks progress toward the next training cycle. Each time a clinician clicks "Accept" on a prediction, the case is added to the learning pool, and the counter increments. Here it reads 12/50, meaning twelve accepted cases toward a threshold of fifty. Two controls support routine operations: Reset Counter clears the tally if the pool should start fresh, and Force Retrain starts a new job immediately when an experiment or urgent update is needed.
- Two scores in the Model Drift section summarise that comparison. PSI = 0.1587 and JSD = 0.0191 are below their respective limits (0.20 and 0.10), so no action is needed. The exact windows are printed, covering the recent period from 2025-09-23 to 2025-09-30, the baseline from 2025-08-24 to 2025-09-23, and a "last checked" time, so anyone reviewing the numbers knows precisely what data were used. The bar chart underneath shows the percentage for each class in both periods.

Over the last ~20 training cycles, as shown in Fig. 4, the validation accuracy chart and model versions display a steady trend in the mid-90% range across recent model versions, with the best model achieving a validation accuracy of ~94.6%. That pattern suggests the auto-retraining loop is doing its job: it brings in newly accepted, clinician-verified cases, updates the model, and preserves or nudges up validation accuracy without the regressions that can appear in ad-hoc updates. In practical terms, the loop appears to maintain stable performance as the data changes and the dataset grows. The absence of sharp drops implies the guardrails (cooldown, locks, best-model selection) help avoid harmful deployments, while the gentle upward trend hints that curated examples are adding real value.

Each model version is released with a full per-class evaluation metrics report rather than a single accuracy figure. For every BI-RADS category, the evaluation records the four confusion counts: true positives (TP), false positives (FP), true negatives (TN), and false negatives (FN), as well as the class support (TP + FN). From these, standard measures are computed: sensitivity/recall TP/(TP + FN), specificity TN/(TN + FP), precision TP/(TP + FP), F1-score 2PR/(P + R), and one-vs-rest AUC using the model's probability outputs.

Table 1 provides an example of a full evaluation metrics report for a single model: birads1 performs strongly with high sensitivity (0.9786), solid specificity (0.9349), and an AUC of 0.9918 over 374 cases; birads2 is perfect on this split (all metrics = 1.0000) across 52 cases; birads3 is acceptable but clearly weaker (sensitivity 0.8148, precision 0.7174, F1 0.7630), pointing to more borderline presentations; birads4 has high specificity (0.9903) but low sensitivity (0.1786), suggesting many misses and a need for more training data or threshold tuning; birads5 also suffers from very small support with sensitivity 0.3750 despite high specificity 0.9832, indicating unstable estimates that will benefit from targeted data collection. Reporting these per-class evaluation metrics for each retrain makes progress measurable, highlights class imbalance (especially BI-RADS 4–5), and guides where the accepted-case pipeline should focus to raise clinical usefulness.

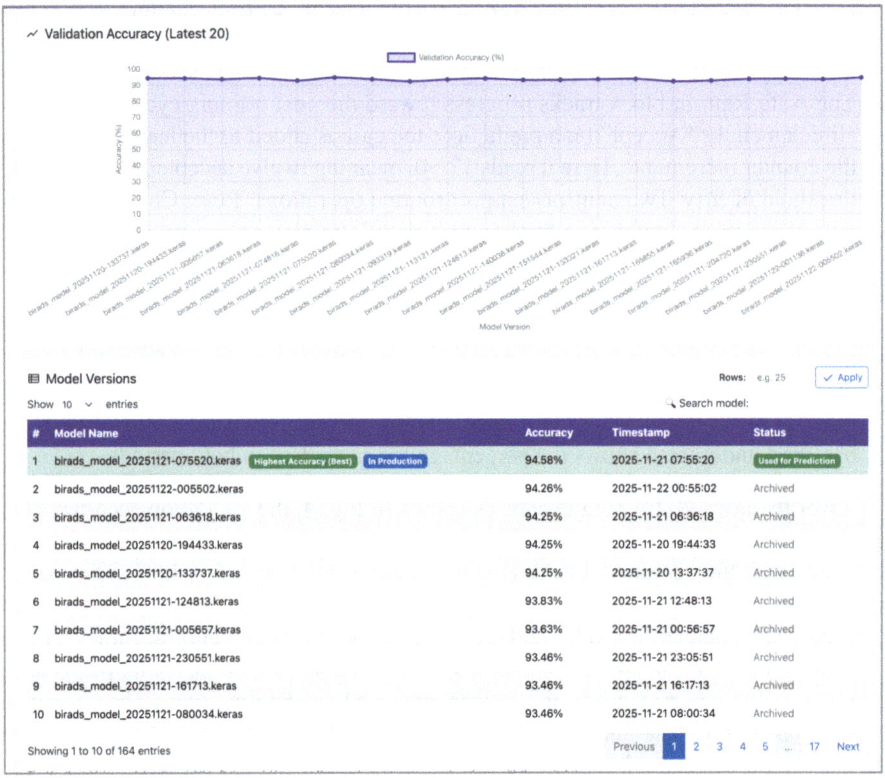

Fig. 4. Validation accuracy and model versions

Table 1. Evaluation metrics report

Class	TP	FP	TN	FN	Support	Sensitivity	Specificity	Precision	Recall	F1	AUC
birads1	366	11	158	8	374	0.9786	0.9349	0.9708	0.9786	0.9747	0.9918
birads2	52	0	491	0	52	1.0000	1.0000	1.0000	1.0000	1.0000	1.0000
birads3	66	26	436	15	81	0.8148	0.9437	0.7174	0.8148	0.7630	0.9585
birads4	5	5	510	23	28	0.1786	0.9903	0.5000	0.1786	0.2632	0.9202
birads5	3	9	526	5	8	0.3750	0.9832	0.2500	0.3750	0.3000	0.9586

5 Conclusion

The proposed MLOps framework delivers real-time BI-RADS classification with predictable speed and controlled change. In production, requests complete well under the service target of p95 < 300 ms, with typical behaviour around p50 $\approx$ 33 ms and p95 $\approx$ 60–80 ms, enabling on-the-spot decisions during reading sessions. Beyond speed, continuous checks of class-mix drift (PSI and JSD) and an acceptance-gated retraining loop ensure the model remains aligned with current data, preventing quality degradation between infrequent manual updates. Across approximately twenty successive retraining cycles, validation accuracy remained in the mid-90% range, peaking at approximately 94.6%. This steady profile, with a slight upward drift, indicates that adding clinician-verified cases through the auto-retraining loop sustains and modestly improves model performance. The lack of sharp drops suggests the safeguards (cooldown, locks, best-model selection) effectively prevent harmful updates. Overall, the pipeline adapts to new data while maintaining a reliable level of clinical usefulness. Compared to a conventional machine-learning setup that focuses on offline accuracy and occasional batch retraining [13–15], this design reduces the time from shift detection to model refresh from days or weeks to minutes or hours, maintains stable performance as data evolves, and provides clear traceability and explanations for every prediction.

Acknowledgement. This study was funded by the Ministry of Higher Education Malaysia (MOHE) through the Fundamental Research Grant Scheme (FRGS) with project reference code: FRGS/1/2023/ICT06/UNISZA/02/1. Thanks to the Centre for Research Excellence & Incubation Management (CREIM) of Universiti Sultan Zainal Abidin (UniSZA).

References

1. Choi, J.S.: Breast imaging reporting and data system (BI-RADS): advantages and limitations. J. Korean Soc. Radiol. **84**(1), 3–14 (2023)
2. Deho, O.B., Liu, L., Li, J., Liu, J., Zhan, C., Joksimovic, S.: When the past != the future: assessing the impact of dataset drift on the fairness of learning analytics models. IEEE Trans. Learn. Technol. **17**, 1007–1020 (2024)
3. Barazi, H., Gunduru, M.: Mammography BI-RADS grading. In: StatPearls [Internet]. StatPearls Publishing (2023)
4. Oza, P., Sharma, P., Patel, S., Bruno, A.: A bottom-up review of image analysis methods for suspicious region detection in mammograms. J. Imaging **7**(9), 190 (2021)
5. Bhowmik, A., et al.: Automated triage of screening breast MRI examinations in high-risk women using an ensemble deep learning model. Investig. Radiol. **58**(10), 710–719 (2023)
6. Garg, P., Krishna, M., Kulkarni, P., Horne, D., Salgia, R., Singhal, S.S.: Machine learning models for predicting gynecological cancers: advances, challenges, and future directions. Cancers **17**(17), 2799 (2025)
7. Ahmad Sukri, N.F.A., Wan Hamzah, W.M.A.F., Yusof, M.K., Ismail, I., Yusoff, H.M., Yacob, A.: A systematic literature review on machine learning in healthcare prediction. Int. J. Online Biomed. Eng. **21**(6) (2025)
8. Ng, M.Y., Youssef, A., Pillai, M., Shah, V., Hernandez-Boussard, T.: Scaling equitable artificial intelligence in healthcare with machine learning operations. BMJ Health Care Inform. **31**(1), e101101 (2024)

9. Hewage, N., Meedeniya, D.: Machine learning operations: a survey on MLOps tool support (2022). arXiv preprint arXiv:2202.10169
10. Berberi, L., et al.: Machine learning operations landscape: platforms and tools. Artif. Intell. Rev. **58**(6), 167 (2025)
11. Garg, S.: Robust machine learning: detection, evaluation and adaptation under distribution shift. Doctoral dissertation, Carnegie Mellon University (2024)
12. Chitnis, A., Tewari, S.: Detecting data drift and ensuring observability with machine learning automation. IRE J. **6**(8), 301–310 (2022)
13. Mahadevan, A., Mathioudakis, M.: Cost-effective retraining of machine learning models (2023). arXiv preprint arXiv:2310.04216
14. Zhou, Z., Peng, A., Li, Q., Levine, S., Kumar, A.: Efficient online reinforcement learning fine-tuning need not retain offline data (2024). arXiv preprint arXiv:2412.07762
15. Bartz-Beielstein, T., Hans, L.: An experimental comparison of batch and online machine learning algorithms. In: Online Machine Learning: A Practical Guide with Examples in Python, pp. 105–124. Springer, Singapore (2024)

Human Anomalous Activity Detection System Based On Infrared Grid Sensor

Ololade Babatunde[1(✉)] and Raqibul Hasan[2]

[1] Department of Computer Engineering, Izmir Institute of Technology, Izmir, Turkey
ololadebabatunde@iyte.edu.tr
[2] Center for Computational and Data Sciences, Independent University, Bangladesh, Dhaka, Bangladesh
raqibul.hasan@iub.edu.bd

Abstract. This study addresses the difficulties of detecting anomalous human activities using low-resolution infrared grid sensor data. Traditional anomaly detection systems either require abundant labeled anomalies or rely on unsupervised reconstruction-based methods that often fail to capture complex activity-level deviations. We conducted three experiments: (1) a *supervised* convolutional neural network baseline trained on activity-level labels while varying the fraction of anomalous training data from 10% to 70%; (2) an *unsupervised* autoencoder trained solely on normal data with reconstruction-error thresholding; and (3) a *semi-supervised* pipeline that fuses autoencoder latent embeddings with raw-data K-means clustering centroid-distance features and a shallow feedforward neural network classifier. We validate our approaches on the publicly available *Infrared Human Activity Recognition Dataset (Coventry-2018)* and observe that while unsupervised autoencoders perform well on distributional anomalies, their effectiveness diminishes for activity-level anomalies. In contrast, the semi-supervised strategy consistently detects novel activities without extensive anomaly labels, demonstrating scalable and label-efficient anomaly detection for real-world elderly surveillance systems.

Keywords: Human Activity Recognition · Anomaly Detection · Semi-Supervised Learning

1 Introduction

Anomalous human activity detection is critical for applications, such as elderly surveillance systems, where timely identification of abnormal behaviors improves safety and responsiveness. Traditional approaches rely on high-resolution video or wearable sensors, raising privacy concerns or requiring active user participation. Low-resolution thermal imaging sensors, such as 8×8 infrared grid arrays, offer a non-intrusive, privacy-preserving alternative that supports passive and continuous monitoring. However, the coarse spatial resolution of such sensors

H.-S. Ma and Y.-W. Chan (Eds.): IC 2026, LNEE 1646, pp. 75–81, 2027.
https://doi.org/10.1007/978-981-92-0779-4_8

poses significant challenges for identifying and differentiating complex human activities, especially when anomalous behaviors are subtle. [1]

This study addresses the problem of detecting anomalous human activities from low-resolution infrared grid sensor data, using the publicly available *Infrared Human Activity Recognition Dataset - Coventry-2018.* We used three complementary modeling paradigms: (1) a *supervised* convolutional neural network (CNN) trained on activity-level labels while varying the fraction of anomalous samples available during training (10%–70%). (2) an *unsupervised* convolutional autoencoder (AE) trained solely on normal sequences, using reconstruction error as an anomaly score, and (3) a *semi-supervised* pipeline that fuses AE latent embeddings with raw-data K-means centroid–distance features and trains a lightweight feedforward neural network (FNN) to exploit limited anomaly labels. Together, these experiments quantify how far unsupervised and semi-supervised methods can go when anomaly labels are scarce, and how supervised performance scales with increased anomaly coverage.

Among the most influential approaches to visual anomaly detection, Hasan et al. learned temporal regularity in videos with an unsupervised reconstruction/prediction objective, showing that deviations from learned normal dynamics expose anomalies without labels [11]. Sultani et al. addressed label scarcity with a weakly supervised multiple-instance learning framework that ranks video segments using only video-level anomaly tags[5]. Georgescu et al. advanced self-supervised, multi-task objectives to boost generalization across scenes [3]. While effective on RGB surveillance benchmarks, these methods typically assume higher spatial resolution and richer visual cues; our focus on 8×8 infrared arrays prioritizes privacy and edge feasibility, where label efficiency and structure-aware representation learning become critical. Rather than proposing new architectures, this work contributes a focused evaluation of supervised, unsupervised, and semi-supervised strategies for 8×8 infrared grids under extreme label scarcity, a setting that remains underexplored despite its practical relevance.

2 Related Works

Early research on anomaly detection in human behavior focused on handcrafted and statistical models. Techniques such as optical flow descriptors, spatio-temporal cubes, sparse coding, and one-class SVMs were widely used to separate abnormal from normal patterns in surveillance data and network logs [1]. Clustering algorithms like K-Means and Local Outlier Factor (LOF) were employed to build unsupervised baselines using reconstruction and distance-based anomaly scores. With the introduction of depth cameras and skeleton tracking, sequence modeling techniques like RNNs, LSTMs, and GRUs gained traction, achieving high recognition accuracy and anomaly detection rates in simulations of elderly care and dementia scenarios [2]. More recent advances include self-supervised, contrastive, and generative formulations designed to improve cross-scene generalization and representation robustness. Despite rich literature, few efforts have addressed the challenge of anomaly detection using privacy-preserving, illumination-invariant infrared grid sensors. Most prior works focus on RGB or depth modalities, rely on synthetic data, or lack principled cross-setting evaluations. In contrast, our study benchmarks convolutional autoencoders, latent clustering, and supervised CNN baselines using the Coventry-2018 Infrared Human Activity Recognition dataset, exploring multiple detection strategies across labeled and unlabeled anomaly conditions with ROC/PR analysis and multi-split evaluation [8]. However, few studies have examined how these approaches behave on privacy-preserving low-resolution infrared grids, especially under severe anomaly-label imbalance.

3 Materials and Methods

We used the **Infrared Human Activity Recognition Dataset (Coventry 2018)** [9] available from IEEE DataPort. It consists of time-series recordings from low-resolution 8×8 thermal sensors capturing human activities in indoor environments [10]. We focused on the **"Big Area"** scenario from **Sensor1**, positioned at the ceiling for an overhead thermal view. Files in `Sensor1_BigArea_Single` were treated as **normal** (single-person) activities, and those in `Sensor1_BigArea_Double` as **anomalous** (two-person) interactions. Each CSV file records about one second of activity (10 fps), reshaped into $40 \times 64 \times 1$ tensors, and normalized by subtracting 10 and dividing by 12.5.

Each sequence was truncated to 40 frames, reshaped to $40 \times\ 64 \times 1$, and normalized as $x_{\text{norm}} = (x-10)/12.5$. These steps ensured consistent input shapes for convolutional networks and stable training across three different experiments.

Experiment 1 (Supervised CNN Baseline): A 2D CNN was trained on labeled normal and anomalous activity samples. "Left & Right Move" and "Forward-Backward Move" were treated as anomalous, while the remaining six activities were normal. The fraction of anomalous training data was varied from 10% to 70%, matching the number of normal samples. Dropout, L2 weight decay, early stopping, and learning rate scheduling were used to avoid overfitting.

Experiment 2 (Unsupervised AE Thresholding): A convolutional autoencoder was trained using only normal activity data from `Sensor1_BigArea_Single`. The mean squared reconstruction error was used as an anomaly score [11], and the threshold was set as median + 3×STD of validation errors.

Experiment 3 (Semi-Supervised AE + K-Means + FNN): A convolutional autoencoder was trained on normal data only, then K-Means clustering was applied to the raw normal data to obtain centroids. Distances from each sample to these centroids were concatenated with AE latent features, and a shallow FNN was trained on this hybrid representation using limited anomalous samples (10%–70%).

All experiments were evaluated using accuracy, precision, recall, F1-score, ROC-AUC, PR-AUC, and confusion matrices. Models were implemented in TensorFlow and scikit-learn with stratified splits for training, validation, and testing.

4 Results

4.1 Experimental Setup

Training and evaluation were performed locally on the CPU, as the infrared grid sensor dataset is lightweight and does not require GPU acceleration for convergence. All models were evaluated on a held-out test set containing both normal and anomalous sequences. The *unsupervised* autoencoder (Experiment 2) was trained exclusively on normal data, whereas the *supervised* CNN (Experiment 1) and the *semi-supervised* AE+K-means+FNN (Experiment 3) were trained with normal data plus varying fractions of anomalous samples (10%–70%), using balanced mini-batches during training as described above.

4.2 Experiment 1: Supervised CNN with Varied Anomaly Fractions

Experiment 1 served as our supervised baseline. We trained a 2D-CNN classifier on balanced sets of normal and anomalous frames, where the amount of anomalous data in the training set was systematically varied from 10% to 70% of the total available anomalies. For each anomaly fraction, we matched an equal number of normal samples to maintain class balance.

Performance at Low Anomaly Coverage (10%–40%). When trained with just 10%–40% of the anomaly samples, the CNN achieved relatively low F1-scores and accuracy on the test set. Although some generalization occurred, the model struggled to distinguish anomalies from normal behavior reliably, often overfitting to the dominant class in the small training set.

Improvement Beyond 50% Anomalies. As the fraction of labeled anomalies increased beyond 50%, the CNN began to generalize more consistently. At 60% anomaly coverage, both accuracy and F1-score rose significantly, suggesting that sufficient anomaly representation enables the model to learn discriminative features more robustly. Variance across runs also reduced, indicating improved training stability (Fig. 1).

4.3 Experiment 2: Unsupervised Autoencoder Anomaly Detection

In this experiment, we trained a convolutional autoencoder on only normal activity sequences from the `Sensor1_BigArea_Single` folder and tested its ability to identify anomalous sequences from `Sensor1_BigArea_Double` based on reconstruction error. After training, the mean squared error (MSE) between the input and reconstructed output was used as the anomaly score. A threshold was computed from the held-out validation set using the formula: $\tau = \text{median} + 3\sigma$. The model achieved high performance across all metrics: **Accuracy = 0.9961**, **Precision = 1.0000**, **Recall = 0.9952**, and **F1-score = 0.9976**. Out of 258 total test samples, the model correctly classified 257, missing only one anomaly.

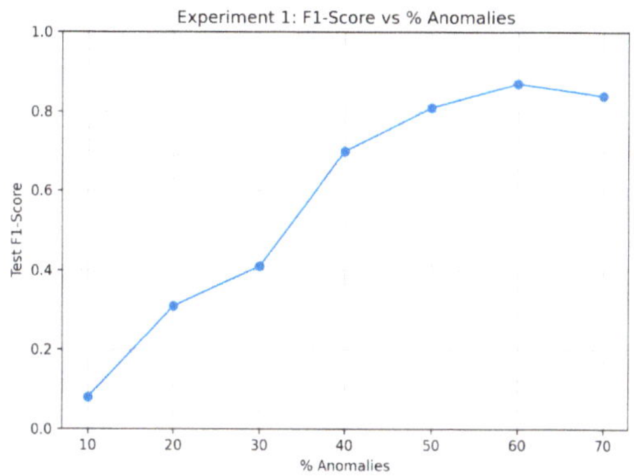

Fig. 1. Experiment 1: F1-score vs. anomaly fraction for the supervised CNN.

4.4 Experiment 3: Hybrid Autoencoder + Raw K-Means + FNN with Varied Anomaly Fractions

In this final experiment, we evaluate our hybrid semi-supervised pipeline (Training a convolutional autoencoder on normals, clustering raw frames with K-means, and concatenating latent + distance features to train an FNN) under the same limited-anomaly regime as Experiment 1. We vary the fraction of anomalous samples in the FNN's training set from 10 % to 70 %, and test on the held-out remainder (anomalies + matching normals).

Results. Even with only 10% anomaly labels, the model achieves a strong F1-score of ≈ 0.95 and accuracy of ≈ 0.96. Both metrics improve steadily with more anomalies, surpassing 0.98 F1 and 0.99 accuracy at 50% anomaly fraction. This consistent monotonic trend highlights the robustness of the hybrid method, and establishes it as a competitive semi-supervised alternative to fully supervised approaches (Fig. 2).

Comparison to Supervised Baseline. Table 1 compares the supervised CNN from Experiment 1 with the semi-supervised hybrid model from Experiment 3 across varying percentages of anomaly labels. Across all anomaly fractions, the hybrid pipeline consistently outperformed the CNN, especially under low-label conditions. At 10% anomalies, the CNN achieved F1 = 0.08 versus 0.9498 for the hybrid model, and even at 60–70% anomalies the hybrid approach remained superior. This confirms that supervised learning degrades sharply when anomaly labels are scarce, while semi-supervised feature learning remains robust.

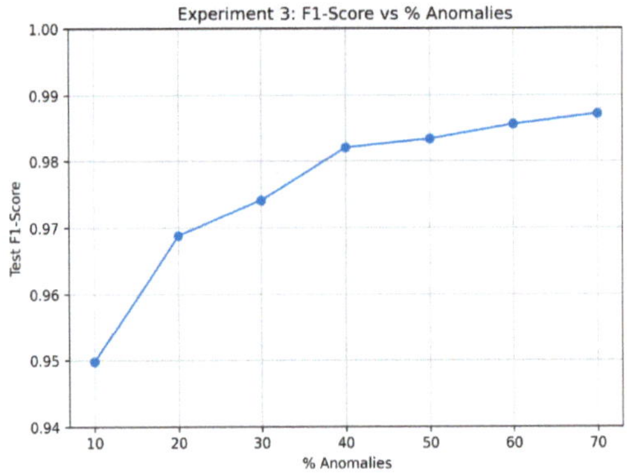

Fig. 2. Experiment 3: F1-score vs. % anomalies in training for the hybrid AE + K-Means + FNN model.

Table 1. F1-Score and Accuracy Comparison Between Experiment 1 (Supervised CNN) and Experiment 3 (Hybrid AE + K-Means + FNN).

% Anomalies	F1 (Exp 1)	Acc (Exp 1)	F1 (Exp 3)	Acc (Exp 3)
10%	0.08	0.40	0.9498	0.9547
20%	0.31	0.52	0.9688	0.9719
30%	0.41	0.57	0.9741	0.9769
40%	0.70	0.74	0.9821	0.9840
50%	0.81	0.80	0.9834	0.9856
60%	0.87	0.86	0.9856	0.9879
70%	0.84	0.86	0.9872	0.9897

5 Conclusion

This study systematically evaluated supervised, unsupervised, and semi-supervised anomaly detection strategies for low-resolution infrared sensor data. The unsupervised AE detects broad distributional shifts but struggles with activity-level anomalies because the 8×8 grid lacks detail and the model learns only dominant normal patterns, limiting generalization to unseen events. The high accuracy reflects the constrained dataset conditions and should not be interpreted as general performance without further validation. The supervised CNN improved with more labeled anomalies but remained dependent on abundant supervision. Because our models operate primarily on frame-level representations, they do not capture the full temporal structure of activities, which limits discrimination of motion-dependent anomalies. In contrast, the hybrid autoencoder-K-means-FNN pipeline consistently achieved strong detection even with minimal labeled data, highlighting the robustness of semi-supervised feature learning for real-world ambient systems. These results reflect a controlled single-sensor, single-scene setting, so broader validation across placements and environments remains necessary. Future work will extend this framework to multi-sensor and temporal models, incorporate contrastive or sequence-aware pretraining, evaluate real-time edge deployment constraints, and assess generalization to unseen anomaly types.

References

1. Dhiman, H., Vishwakarma, S.: A review of techniques for abnormal human activity recognition. Signal Process. Image Commun. **63**, 74–92 (2018)
2. Malhotra, P., Vig, L., Shroff, G., Agarwal, P.: LSTM-based encoder–decoder for multi-sensor anomaly detection. In: ICML Time Series Workshop (2016)
3. Georgescu, R., Irani, A., Penny, W.: Anomaly detection via self-supervised and multi-task learning. In: CVPR, pp. 12779–12788 (2021)
4. Johnson, A., Ramachandran, P.: AnomalyNet: an anomaly detection network for surveillance. Pattern Recognit. Lett. **136**, 265–273 (2020)
5. Sultani, W., Chen, C., Shah, M.: Real-world anomaly detection in surveillance videos. In: CVPR, pp. 6479–6488 (2018)
6. Fahim, M., Sillitti, A.: Anomaly detection and prediction in IoT: a review. IEEE Access **7**, 163834–163858 (2019)
7. Usiade, J., Adeoye, P.: Smart home security and intrusion detection system. In: Proceedings of International Symposium and Embedded Systems, pp. 45–52 (2020)
8. Markou, M., Singh, S.: Novelty detection: a review–Part 1. Signal Process. **83**(12), 2481–2497 (2003)
9. Hassan, M., Wilcox, L.D., Dixit, A., Ghaleb, F., Davis, I., Brusey, J.: Infrared HAR Dataset – Coventry. IEEE Dataport (2018). https://ieee-dataport.org/documents/infrared-human-activity-recognition-dataset-coventry-2018
10. Fu, C., Zhang, Y., Li, H., et al.: Human activity recognition using 8×8 IR sensor arrays. Sensors (2024)
11. Hasan, M., Choi, J., Neumann, J., Roy-Chowdhury, A.K., Davis, L.S.: Learn. temporal regularity in video seqs. In: CVPR (2016)
12. Ruff, L., Vandermeulen, R., Görnitz, N.: Deep one-class classification. In: ICML (2018)

Nomophobia in University Students: A Machine Learning and Data Analysis Approach to Smartphone-Related Anxiety

Carlos Alberto Espinosa-Pinos(✉), Paola Silvana Valencia-Medina, and Camila Alessandra Valarezo-Calero

Universidad Tecnológica Indoamérica, Ambato, Ecuador
carlosespinosa@indoamerica.edu.ec

Abstract. Mental health concerns such as nomophobia a disorder marked by excessive mobile phone dependence and anxiety are on the rise among university students, yet detection strategies remain underexplored. This study used supervised machine learning methods to predict nomophobia risk in 1,484 Ecuadorian students (aged 18–25) using the Nomophobia Questionnaire (NMP-Q) and the DASS-21 Anxiety subscale. Multiple algorithms (Random Forest, SVM, neural networks, XGBoost, decision trees) were compared across different outcome fusions. The Random Forest model with binary classification ("with vs. without nomophobia") showed the strongest performance (F1 = 0.82, Accuracy = 0.72, Precision = 0.89, Recall = 0.76). Variable importance analysis revealed anxiety symptoms, academic context, and technological factors as the top predictors, while standard demographics contributed minimally. Findings indicate that machine learning especially Random Forest with binary fusion offers a robust approach for digital mental health screening and targeted intervention among university students.

Keywords: Anxiety · Machine learning · Nomophobia · University students · Smartphone use

1 Introduction

The rapid expansion of mobile technology has transformed communication, learning, and social interaction in the 21st century. However, it has also fostered new psychological challenges, among which nomophobia the fear or anxiety of being without one's mobile phone stands out [1, 2]. Conceptualized as a situational phobia, nomophobia is marked by emotional, cognitive, and physiological responses when mobile connectivity cannot be maintained [3]. Anxiety symptoms are central to the nomophobia experience. While many studies report a positive association between nomophobia and anxiety in university students, findings are mixed and may depend on specific contextual and psychosocial variables [4]. The role of demographic factors such as age and gender remain controversial, with some evidence suggesting that technological, academic, and emotional variables are stronger determinants of risk [5]. Traditional research has relied

H.-S. Ma and Y.-W. Chan (Eds.): IC 2026, LNEE 1646, pp. 82–87, 2027.
https://doi.org/10.1007/978-981-92-0779-4_9

on correlation and group comparison methods, which may not capture the complexity of nomophobia and its predictors. Machine learning classification models provide a more powerful approach by simultaneously assessing the predictive contribution of multiple variables and uncovering nonlinear patterns [2–7]. However, few studies in Latin America have applied such models to nomophobia, and the relative importance of sociodemographic and psychosocial predictors remains unclear.

This study fills a gap by applying multiple supervised machine learning classifiers to predict nomophobia risk among Ecuadorian university students. Psychological factors (anxiety symptoms), academic variables (study program, schedule), and technological access (internet) were the strongest predictors, while traditional demographics had minimal influence. The Random Forest model with binary classification ("with vs. without nomophobia") achieved the highest accuracy, highlighting machine learning's value in digital mental health screening in higher education contexts.

2 Methodology

2.1 Participants

A total of 1,484 students from two higher education institutions in Ambato, Ecuador, participated in the study. Inclusion criteria comprised: (a) age between 18 and 25 years, (b) current enrollment in undergraduate programs, (c) regular use of a mobile device, and (d) signed informed consent. Exclusion criteria included any self-reported medical or psychiatric diagnosis and current substance use. Participant eligibility was assessed exclusively through self-reported screening items embedded within the online survey. Importantly, no external clinical verification was conducted. We acknowledge that reliance on self-report may result in under-reporting or inaccurate disclosure of pre-existing medical or psychiatric conditions, potentially introducing bias into the sample composition.

2.2 Procedures

A quantitative, predictive, non-experimental design was used. Data were collected online via Microsoft Forms, including informed consent, sociodemographic information, the Nomophobia Questionnaire (NMP-Q), and DASS-21 anxiety subscale. Ethical approval and informed consent were obtained. After cleaning and coding, data were split into 70% training and 30% test sets. All predictors were standardized or factorized. Exclusion criteria (medical, psychiatric conditions, substance use) were assessed by self-report, without clinical verification, which may have introduced bias in anxiety-related outcomes.

2.3 Instruments and Machine Learning Analysis

Sociodemographic data including age, sex, gender, marital status, socioeconomic status, academic program, study schedule, home location, and internet access were collected. Nomophobia was assessed with the 20-item NMP-Q, showing excellent reliability ($\alpha = 0.957$; CFI = 0.934; SRMR = 0.038; [2]), and anxiety with the DASS-21 subscale [9, 10, 11]. Supervised classification algorithms (Random Forest, SVM, neural networks,

XGBoost, decision trees) were compared using binary and multiclass outcomes. Model evaluation included accuracy, recall, precision, and F1-score; Random Forest with binary outcomes showed the best performance (F1 = 0.82, Accuracy = 0.72, Precision = 0.89, Recall = 0.76). Variable importance highlighted anxiety, age, academic program, study schedule, internet access, and socioeconomic status as main predictors, with gender and marital status showing minimal contribution. Analyses were conducted in RStudio with specialized packages.

3 Results

3.1 Model Comparison and Performance

Several machine learning classifiers were tested to predict nomophobia risk using a binary outcome. Table 1 and Fig. 1 compare the accuracy, recall, precision, and F1-score of each model, showing that the Random Forest with binary fusion performed best.

Table 1. Comparative performance of classification models

Fusion Scheme	Model	Accuracy	Recall	Precision	F1
2 Classes (with vs. without nomophobia)	Random Forest	0.72	0.76	0.89	0.82
2 Classes (with vs. without nomophobia)	XGBoost	0.70	0.74	0.88	0.81
2 Classes (with vs. without nomophobia)	Neural Network	0.68	0.73	0.86	0.79
2 Classes (severe vs. rest)	Random Forest	0.84	0.55	0.40	0.38
3 Classes (intermediate)	Random Forest	0.55	0.40	0.38	0.37

Note. Multiclass and risk-based fusions scored lower, confirming the advantage of the binary approach

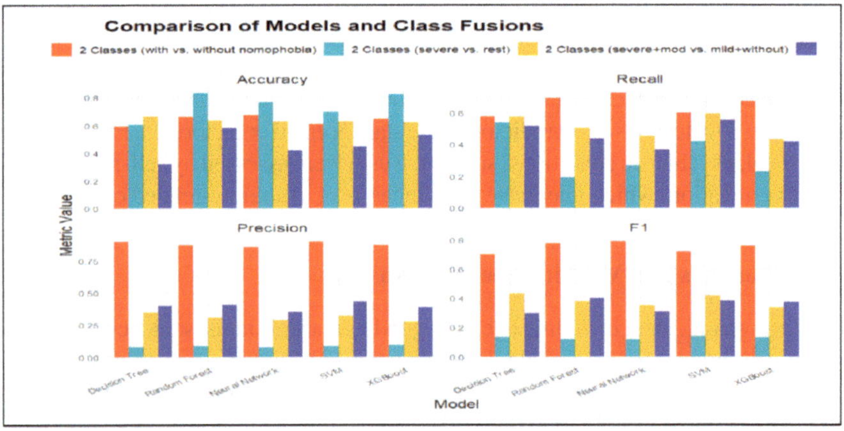

Fig. 1. The Random Forest binary model shows the best balance of accuracy, recall, precision, and F1-score.

3.2 Variable Importance in the Best Model

To determine which features contributed most to the classification of nomophobia risk, we analyzed variable importance scores from the best-performing Random Forest model (see Fig. 2). Figure 2 illustrates that psychological and contextual variables dominated the prediction.

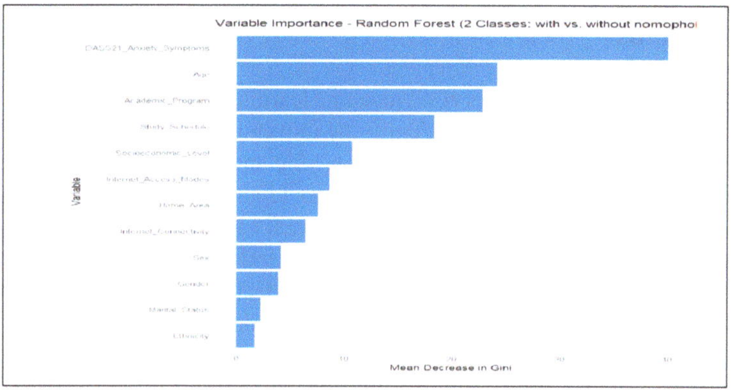

Fig. 2. Anxiety, academic, and technological factors were the main predictors of nomophobia; demographics mattered little.

Anxiety symptoms were the most influential predictor of nomophobia, with age, academic program, study schedule, internet access, and socioeconomic status also relevant. Students with higher anxiety, younger age, specific academic backgrounds, and certain patterns of internet use or socioeconomic status were at greater risk. In contrast, demographic variables such as sex, gender, and marital status contributed minimally, highlighting the greater importance of psychological, educational, and technological factors for accurate classification.

4 Discussion

This study shows that machine learning classifiers, particularly Random Forest, accurately identify students at risk of nomophobia by integrating psychological, academic, technological, and demographic factors. Consistent with earlier research [2–7], Random Forest delivered the best performance. Anxiety symptoms measured with DASS-21 were the strongest predictor of risk, supporting literature linking mobile dependence to elevated anxiety and distress [4–10]. Age and academic program also stood out younger students and those in demanding fields faced higher nomophobia risk, reflecting the psychosocial demands of academic transitions and professional uncertainty [12]. Additionally, variables like study schedule, internet access, and socioeconomic status were important contributors, highlighting the relevance of academic and technological context in digital mental health prediction [4].

Traditional demographic factors, such as gender, sex, and marital status, contributed minimally to our classification model, supporting that nomophobia is more closely associated with emotional and contextual variables than static sociodemographics [5]. Although we observed a strong link between anxiety and nomophobia, our cross-sectional design prevents causal interpretation and directionality; further longitudinal and experimental studies are needed. Importantly, the strength of machine learning lies in its ability to quantify multiple predictors, emphasizing the importance of prioritizing emotional health and digital context in student screening and intervention strategies [8].

5 Conclusions

This study demonstrates that supervised machine learning, particularly the Random Forest algorithm, is highly effective for identifying university students at risk of nomophobia. Psychological and contextual factors especially anxiety symptoms, age, academic program, study schedule, socioeconomic status, and internet access emerged as key predictors, while traditional demographics had minimal impact. These results underscore the importance of focusing on emotional well-being and digital contexts in student mental health screening. However, our cross-sectional design limits causal interpretation, and reliance on self-reported clinical status may have introduced undetected comorbidity. Future research should use longitudinal methods and diverse samples to clarify these relationships and further improve predictive models.

6 Limitations and Future Directions

This study has several limitations that merit attention. First, the cross-sectional design prevents any inference about causality between nomophobia and anxiety; the directionality of their association remains unresolved and requires future longitudinal research. Second, all exclusion criteria and psychological measures relied solely on self-reported online survey responses, without independent clinical verification. This may have allowed under-reporting of medical or psychiatric conditions, introducing potential bias in anxiety levels and overall findings. Third, the sample was restricted to students from two universities in Ambato, Ecuador, limiting the generalizability of results. Future research should employ longitudinal designs and broader recruitment, as well as incorporate objective behavioral metrics, to strengthen model validity and clarify the mechanisms linking nomophobia and anxiety.

Acknowledgment. This study is part of a larger research project on digital dependence and student mental health. The authors thank the institutions in Ambato, Ecuador, for their valuable support and sponsorship.

References

1. Al Maghaireh, D.F., Shawish, N.S., Abu Kamel, A.M., Kawafha, M.: Acute Nomophobia and its psychological correlates in adolescents: an explanatory sequential mixed-methods approach. J. Multidiscip. Healthc. **18**, 1445–1460 (2025). https://doi.org/10.2147/JMDH.S505535;WGROUP:STRING:PUBLICATION

2. Altınel, B., Uyaroğlu, A.K., Ergin, E.: The effect of social appearance anxiety and loneliness on nomophobia levels of young adults. Arch. Psychiatr. Nurs. **50**, 27–32 (2024). https://doi.org/10.1016/J.APNU.2024.03.009
3. Arora, A., Chakraborty, P., Bhatia, M.P.S.: Real time smartphone data for prediction of Nomophobia severity using supervised machine learning. Proceedings of International Conference on Women Researchers in Electronics and Computing, pp. 78–83 (2021). https://doi.org/10.21467/PROCEEDINGS.114.11
4. Aydın, K., Geçikli, F.: A study of the prevalence of nomophobia and its associated variables among university students. SN Soc. Sci. **4**(7), 132 (2024). https://doi.org/10.1007/S43545-024-00927-4
5. Barbayannis, G., Bandari, M., Zheng, X., Baquerizo, H., Pecor, K.W., Ming, X.: Academic stress and mental well-being in college students: correlations, affected groups, and COVID-19. Front. Psychol. **13**, 886344 (2022). https://doi.org/10.3389/FPSYG.2022.886344/FULL
6. Espinosa-Pinos, C.A., Acosta-Pérez, P.B., Valarezo-Calero, C.A.: Applying classification techniques in machine learning to predict job satisfaction of university professors: a sociodemographic and occupational perspective. Computers **13**(12), 344 (2024). https://doi.org/10.3390/COMPUTERS13120344
7. Guerra Ayala, M.J., Alegre de la Rosa, O.M., Chambi Catacora, M.A. del P., Vargas Onofre, E., Cari Checa, E., Díaz Flores, D.: Nomophobia, phubbing, and deficient sleep patterns in college students. Front. Educ. **9**, 1421162 (2024). https://doi.org/10.3389/FEDUC.2024.1421162/BIBTEX
8. Laranjeira, C., Querido, A., Sousa, P., Dixe, M.A.: Assessment and psychometric properties of the 21-Item Depression Anxiety Stress Scale (DASS-21) among Portuguese Higher Education Students during the COVID-19 Pandemic. Eur. J. Invest. Health Psychol. Educ. **13**(11), 2546–2560 (2023). https://doi.org/10.3390/EJIHPE13110177
9. Osman, A., Wong, J.L., Bagge, C.L., Freedenthal, S., Gutierrez, P.M., Lozano, G.: The Depression Anxiety Stress Scales 21 (DASS-21): further examination of dimensions, scale reliability, and correlates. J. Clin. Psychol. **68**(12), 1322–1338 (2012). https://doi.org/10.1002/JCLP.21908
10. Rajguru, A.J., Mishra, A.K., Bhargava, R., Sarkar, S., Balhara, Y.P.S.: Exploring risk factors and determinants: a scoping review of factors associated with nomophobia. Indian J. Psychiatry **66**(7), 591–602 (2024). https://doi.org/10.4103/INDIANJPSYCHIATRY.INDIANJPSYCHIATRY_244_24
11. Vagka, E., Gnardellis, C., Lagiou, A., Notara, V.: Prevalence and factors related to nomophobia: arising issues among young adults. Eur. J. Invest. Health Psychol. Educ. **13**(8), 1467–1476 (2023). https://doi.org/10.3390/EJIHPE13080107
12. Yildirim, C., Correia, A.P.: Exploring the dimensions of nomophobia: development and validation of a self-reported questionnaire. Comput. Hum. Behav. **49**, 130–137 (2015). https://doi.org/10.1016/J.CHB.2015.02.059

Comparative Machine Learning Assessment of Cognitive, Social, and Emotional Skills in University Students Using Tree-Based and Kernel-Based Models

Carlos Alberto Espinosa-Pinos(✉), Jorge Luis Cisneros-Bedón, Noemí Viviana Rodriguez-Ortiz, Daniela Jaramillo, and Alejandro Llerena

Universidad Tecnológica Indoamérica, Ambato, Ecuador
carlosespinosa@indoamerica.edu.ec

Abstract. This study applied machine learning models to predict university students' academic performance based on cognitive, social, and emotional skills, using data from 1,000 students in Nursing, Medicine, and Psychology. Multiple predictors—including sociodemographics, academic semester, emotional intelligence, social skills, and cognitive flexibility—were analyzed. Among five classifiers tested (decision tree, random forest, k-nearest neighbors, support vector machine, gradient boosting), random forest achieved the highest accuracy (0.729) and F1-score (0.823), with gradient boosting and SVM also performing strongly. Key predictors were academic semester, emotional intelligence, age, and social skills. The findings demonstrate the value of automated, multidimensional screening to support early academic intervention, and underscore the importance of psychosocial and cognitive variables in predictive analytics for higher education.

Keywords: Academic performance · Machine learning · Emotional intelligence · Predictive modeling

1 Introduction

Education is a fundamental component of human development, but evaluating student quality solely by grades is limited because emotional intelligence, cognitive abilities, and interpersonal relationships also influence performance, as highlighted by Martínez (2023). Research by Talavera [1] shows that students with better academic outcomes exhibit greater emotional and social well-being, whereas those with lower performance experience more depression and demotivation. Socioemotional learning is therefore crucial; Taveras [2] emphasizes that education should develop not just memory but interpretation and application of knowledge in varied contexts. At the university level, Liu [3] point out that performance is shaped by cognitive, social, and emotional competencies, including memory, attention, and teamwork. Parker [4] assert that emotional intelligence particularly self-regulation and stress management is key to academic success. Finally, Lopes and Salovey [5, 6] demonstrate that programs boosting socioemotional skills directly enhance academic achievement and well-being, supporting interventions that increase motivation, self-esteem, and resilience.

H.-S. Ma and Y.-W. Chan (Eds.): IC 2026, LNEE 1646, pp. 88–93, 2027.
https://doi.org/10.1007/978-981-92-0779-4_10

2 Methodology

2.1 Study Design, Sample and Data Collection Instrument

This cross-sectional study included 1000 university students from Ecuador enrolled in Clinical Psychology and Nursing programs, spanning semesters one to six. Participants were selected using convenience sampling, prioritizing accessibility and willingness to participate, as recommended by Hernández [7]. Inclusion criteria involved enrollment in either program, age between 18–28 years, and signed informed consent; exclusion criteria comprised incomplete or invalid instrument responses and absence of consent. Three validated psychometric instruments were used: the Bar-On Emotional Quotient Inventory (emotional intelligence), the EHS Social Skills Scale (social behaviors), and the Changes Cognitive Flexibility Test (adaptability). Data were collected via computer-based testing in supervised lab sessions. After descriptive statistics, machine learning classification methods including decision trees, support vector machines, random forests, k-nearest neighbors, and gradient boosting were applied to predict academic performance from cognitive, social, and emotional variables. Model evaluation used accuracy, recall, precision, and F1-score, as recommended for robust comparisons. Ethical approval was obtained, ensuring anonymity and voluntary participation throughout. See Table 1.

Model performance was assessed using an 80/20 train-test split and five-fold cross-validation on the training data. Hyperparameters were optimized through grid search, and the best configurations based on cross-validation accuracy were evaluated on the test set.

Table 1. Descriptive characters of the sample (N = 1000).

variable	Categories / description
Degree program	Nursing 329 (32.9%); Medicine 360 (36%); Psychology 311 (31.1%)
Age (years)	18 49 (4.9%); 19 73 (7.3%); 20 128 (12.8%); 21 145 (14.5%); 22 173 (17.3%); 23 157 (15.7%); 24 138 (13.8%); 25 77 (7.7%); 26 41 (4.1%); 27 14 (1.4%); 28 5 (0.5%)
Motivation	High 302 (30.2%); Low 185 (18.5%); Average 513 (51.3%)
Empathy	High 302 (30.2%); Low 205 (20.5%); Average 493 (49.3%)
Assertiveness	High 272 (27.2%); Low 256 (25.6%); Medium 472 (47.2%)
Communication Skills	Good 445 (44.5%); Deficient 122 (12.2%); Fair 433 (43.3%)
Social Skills Global (score)	Mean 500.5 (SD 288.82)
Self-awareness	High 249 (24.9%); Low 175 (17.5%); Average 576 (57.6%)
Self-regulation	High 308 (30.8%); Low 214 (21.4%); Average 478 (47.8%)
Cognitive Flexibility	High 324 (32.4%); Low 192 (19.2%); Average 484 (48.4%)
Flexibility Cognitive Global	High 354 (35.4%); Low 149 (14.9%); Average 497 (49.7%)
Emotional Intelligence Global	Good 333 (33.3%); Developed 407 (40.7%); Needs Improvement 260 (26%)

(continued)

Table 1. (*continued*)

variable	Categories / description
Study shift	Morning 328 (32.8%); Night 368 (36.8%); Afternoon 304 (30.4%)
Semester	1st 114 (11.4%); 2nd 108 (10.8%); 3rd 81 (8.1%); 4th 90 (9%); 5th 103 (10.3%); 6th 97 (9.7%); 7th 98 (9.8%); 8th 96 (9.6%); 9th 104 (10.4%); 10th 109 (10.9%)
Gender	Female 507 (50.7%); Male 493 (49.3%)
Academic Performance 1	Passing 679 (67.9%); Not passing 321 (32.1%)

1Variable to predict.

2.2 Descriptive Analysis

The sample comprised 1,000 university students from Nursing, Medicine, and Psychology, with balanced gender distribution (51% female, 49% male), a mean age of 22, and representation across all semesters and study shifts. Most students (68%) passed their courses. Psychosocial variables including motivation, empathy, assertiveness, self-awareness, self-regulation, and cognitive flexibility were predominantly average, with fewer students at high or low levels. Social skills and communication abilities showed variability but were generally favorable, while emotional intelligence was mostly developed or good, with 26% needing improvement. These characteristics reflect a heterogeneous and robust cohort, supporting the reliability of subsequent machine learning analyses. Model evaluation used an 80/20 train-test split and five-fold cross-validation to ensure robust, unbiased results.

3 Results

Five machine learning models were compared for predicting academic performance. Random Forest achieved the highest accuracy (0.729) and F1-score (0.823), with Gradient Boosting and SVM also performing well. Decision Trees showed solid results, while K-Nearest Neighbors performed lower. Notably, SVM and Random Forest attained the highest recall, effectively identifying students who succeeded academically. (See Table 2 and Fig. 1).

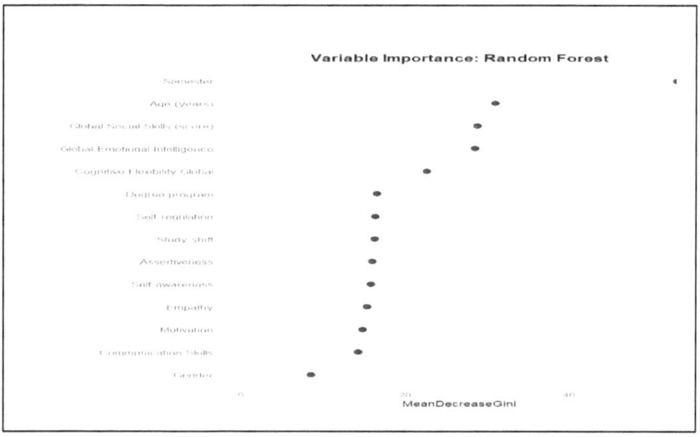

Fig. 1. Variable Importance: Random Forest

Variable importance from the Random Forest model identified academic semester (MeanDecreaseGini = 53.1), age (30.9), global social skills (28.7), and emotional intelligence (28.5) as the top predictors. Other relevant variables included cognitive flexibility, self-awareness, study shift, and self-regulation. Combining sociodemographic and psychosocial features improved model performance compared to using either set alone. These results show that multidimensional models incorporating cognitive, social, and emotional factors can accurately predict academic outcomes. Automated assessment tools are supported for identifying at-risk students and enhancing interventions. All classification metrics accuracy, recall, precision, and F1-score are based on models with optimal hyperparameters (via cross-validated grid search) and were evaluated on an independent test set to ensure generalizability.

Table 2. Classification metrics for prediction.

Model	Accuracy	Recall	Precission	F1
Decision Tree	0.722	0.862	0.761	0.808
Random Forest	0.729	0.926	0.740	0.823
K-Nearest Neighbors	0.585	0.709	0.689	0.699
SVM 1	0.696	0.975	0.697	0.813
Gradient Boosting	0.726	0.911	0.743	0.819

1Variable to predict.

The comparative analysis shows that tree-based models, particularly Random Forest and Gradient Boosting, outperform kernel-based methods like SVM and k-NN in this context. Tree-based algorithms achieved higher accuracy and F1-scores, likely due to their ability to capture complex nonlinear interactions among psychosocial, cognitive, and sociodemographic variables. In contrast, kernel-based models require extensive tuning and may be less effective with high-dimensional, multicollinear data, whereas tree-based models naturally handle such complexities, resulting in better predictive performance for multidimensional educational datasets.

4 Discussion

Machine learning models accurately predicted academic performance in university students, with Random Forest and Gradient Boosting showing the best accuracy and F1-scores, and Support Vector Machines and Random Forest excelling in recall [8, 9]. Key predictors identified were academic semester, emotional intelligence, age, and social skills. The inclusion of psychosocial, cognitive, and sociodemographic variables enhanced predictive accuracy beyond academic-only models (Fig. 1). These results support the integration of multidimensional student data and automated methods for early identification of at-risk students and effective educational interventions (see Table 2).

Several limitations should be noted. Convenience sampling from Nursing, Medicine, and Psychology programs at one Ecuadorian university limits generalizability, and predictor rankings may reflect sample-specific factors. Caution is advised when applying these findings elsewhere. Future research should validate these models in additional disciplines and settings. The cross-sectional design prevents assessment of changes in psychosocial and cognitive skills over time; since emotional intelligence and social skills are modifiable, longitudinal studies are needed to evaluate intervention impact and support personalized strategies [9, 10].

The sample included a higher proportion of 'Passed' (67.9%) than 'Not Passed' (32.1%) students, which may have affected certain performance metrics, especially precision for the 'Not Passed' group. Future research should address class imbalance using resampling and additional metrics such as recall, F1-score, and ROC-AUC for a more accurate evaluation. Also, the costs of errors are relevant: false negatives may leave at-risk students without support, while false positives can misallocate resources. Incorporating cost-sensitive approaches and adjusting thresholds can help optimize support programs and enhance the impact of predictive analytics for timely student interventions [10, 11].

5 Conclusions

This study demonstrates that machine learning models especially random forest and gradient boosting are effective for multidimensional assessment of academic performance in university students. By integrating socio-demographic, cognitive, emotional, and social skills variables, predictive accuracy is notably improved, with academic semester, emotional intelligence, age, and social skills emerging as the most influential predictors. These results support the use of automated tools for early identification of at-risk students and the design of targeted educational interventions. However, generalizability is limited by the sampling strategy and specific institutional context. Future research should replicate these analyses in broader academic and institutional settings, employ longitudinal designs to track evolving psychosocial competencies, and explore advanced techniques such as deep learning and ensemble models to further strengthen predictive power. The rigorous methodology comprising careful data partitioning, cross-validation, and systematic hyperparameter optimization e hances the credibility and reproducibility of the findings, facilitating future implementation of automated assessment frameworks in diverse higher education environments.

Acknowledgment. This study forms part of a broader initiative to assess cognitive, social, and emotional skills in university students. The authors gratefully acknowledge the support of Universidad Tecnológica Indoamérica.

References

1. Martínez-Rodríguez, A., Ferreira, C.: Relationship between academic performance and emotional intelligence in undergraduate and master's degree students at the University of León. Revista Complutense de Educación **34**(4), 795–807 (2023). https://doi.org/10.5209/rced.80128
2. Repetto Talavera, E., Peña Garrido, M.: Socioemotional competencies as a factor of quality in education. REICE **8**(5), 82–95 (2010). https://www.redalyc.org/pdf/551/55119084006.pdf
3. Cao, J., Dai, T., Dong, H., Chen, J., Fan, Y.: Research on the mechanism of academic stress on occupational burnout in Chinese universities. In: Scientific Reports (Vol. 14, Issue 1). Springer Science and Business Media LLC (2024). https://doi.org/10.1038/s41598-024-62984-2
4. Taveras, L.: Cognitive and affective variables predictive of academic performance of university students. Int. J. Educ. Res. Innovation **14** (2021). https://www.upo.es/revistas/index.php/IJERI/article/view/6189/6426
5. Liu, W., Fan, J., Zhao, J.: Cognitive skills and academic performance: the role of attention and memory in higher education. J. Educ. Psychol. **113**(4), 679–695 (2021)
6. Parker, J.D.A., Saklofske, D.H., Keefer, K.V.: Emotional intelligence in education: integrating research with practice. Springer (2018)
7. Hernández, R., Fernández, C., Baptista, P.: Metodología de la investigación. McGraw-Hill (2014)
8. Hastie, T., Tibshirani, R., Friedman, J.: The elements of statistical learning: data mining, inference, and prediction (2nd ed.). Springer (2009). https://doi.org/10.1007/978-0-387-84858-7
9. Sokolova, M., Lapalme, G.: A systematic analysis of performance measures for classification tasks. Inf. Process. Manage. **45**(4), 427–437 (2009). https://doi.org/10.1016/j.ipm.2009.03.002
10. Saito, T., Rehmsmeier, M.: The precision-recall plot is more informative than the ROC plot when evaluating binary classifiers on imbalanced datasets. PLoS ONE **10**(3), e0118432 (2015). https://doi.org/10.1371/journal.pone.0118432
11. Espinosa-Pinos, C.A., Acosta-Pérez, P.B., Valarezo-Calero, C.A.: Applying classification techniques in machine learning to predict job satisfaction of university professors: a sociodemographic and occupational perspective. Computers **13**(12), 344 (2024). https://doi.org/10.3390/computers13120344

Conceptual Design of the Indigenous Knowledge Management System Cycle of Papuan Indigenous Communities

Melkior N. N. Sitokdana(✉), Apol Pribadi Subriadi, and Reny Nadlifatin

Sepuluh Nopember Institute of Technology, Surabaya, Indonesia
sitokdanamelkior@gmail.com

Abstract. The existence of indigenous knowledge (IK) is proven to support sustainable development in various parts of the world, including in Papua, Indonesia. However, the development of globalization and modernization marked by the dominance of western knowledge, transmigration and capital-oriented development has caused indigenous knowledge to be marginalized and no longer passed on to the younger generation. Therefore, strategic and collaborative efforts are needed to save this priceless heritage. One of the efforts made is to develop a conceptual model of the indigenous knowledge management system (IKMS) cycle that can assist stakeholders in Papua in identifying, documenting and utilizing indigenous knowledge for various development purposes. Based on the literature review and the researcher's emic experience, eight elements of the indigenous knowledge management system cycle were formulated, namely initiation, consent, collection, codification, validation, storage, utilization and evaluation. This cycle serves as a guide for stakeholders in Papua to manage information technology-based indigenous knowledge ethically and responsibly.

Keywords: Indigenous Knowledge · Knowledge Management System · Papuan Indigenous Communities

1 Introduction

Indigenous knowledge (IK) is knowledge that is formed through collective experiences, human relations with nature, and cultural practices that are passed down from generation to generation in a community [1]. IK is passed down through oral traditions, cultural symbols, traditional rituals, and collective life experiences that bind indigenous communities in a system of values, norms and responsibilities towards customary lands [2]. Indigenous knowledge also reflects cultural identity and guidelines in natural resource management, conflict management, disaster management, and the formation of social structures of indigenous communities [3–5]. The existence of indigenous knowledge has been proven to support sustainable development in various parts of the world, especially in the agriculture, forestry, tourism, marine and climate change mitigation sectors [6–8]. However, globalization and modernization, characterized by the dominance of western science and technology, capital-oriented development, and population migration from

H.-S. Ma and Y.-W. Chan (Eds.): IC 2026, LNEE 1646, pp. 94–109, 2027.
https://doi.org/10.1007/978-981-92-0779-4_11

one region to another, have caused indigenous knowledge to be eroded, forgotten, and even no longer passed on to the younger generation [9, 10]. The presence of modernization is often not in line with the efforts of indigenous communities to maintain indigenous knowledge, because its influence is so strong and cannot be reversed.

This challenge occurs in Indonesia, especially in Papua, which has been recognized by the state as an indigenous territory. This region has hundreds of tribes and languages, as well as ecological and mineral wealth that holds unique indigenous knowledge [11]. Its existence not only serves as knowledge and daily practices, but is a major pillar in the structure of society, living space, customary power systems and spiritual relationships between humans and ancestral nature [12]. However, the current reality is that indigenous Papuans face multidimensional pressures that threaten the sustainability of the indigenous knowledge system. Various development projects such as large-scale mining, oil palm plantations, rice fields, timber and marine industries, and infrastructure development are often carried out without a Free, Prior, and Informed Consent mechanism, thus triggering land grabbing, deforestation, and agrarian conflicts [13]. Transmigration, social fragmentation, and weak legal protection of indigenous peoples' rights have also accelerated the degradation of indigenous knowledge. The situation is exacerbated by low cultural literacy, poverty and educational inequality hampering the regeneration of knowledge holders [14, 15]. Related to this, the Indonesian government has actually responded through affirmative policies such as Law Number 2 of 2021 concerning Papua Special Autonomy, which emphasizes the importance of affirmation action for indigenous people and development must be based on customary territories, ecological zones, and strengthening customary institutions [16]. However, the implementation is still minimal in internalizing indigenous knowledge into regulations, development planning and policy making. As a result, development is slow and there is easy resistance to development and investment in Papua [17].

Amidst these challenges, on the other hand, information technology presents a new opportunity to help indigenous peoples and stakeholders to document, store and utilize indigenous knowledge. A number of studies show that digitization has been used by indigenous communities to preserve local languages, folklore, agricultural systems, and local cultural expressions [18–20]. Technology also plays an important role in overcoming geographic isolation, strengthening access to local produce markets, and expanding advocacy networks [21]. However, without a systematic framework that respects customary norms and collective rights, technology risks reproducing inequality and knowledge exploitation. In this context, a conceptual approach is needed that is able to bridge local knowledge systems with modern management systems in an ethical, participatory and contextual manner. One strategic approach is the development of an Indigenous Knowledge Management System (IKMS), which is an adaptation of the modern Knowledge Management System (KMS) framework that is tailored to the dynamics, social structure and local values of indigenous peoples. IKMS is designed not only as a technical instrument for data and information management, but as a socio-cultural system that allows indigenous peoples to control the process of identification, documentation and utilization of knowledge.

Based on the literature study, the IKMS model in the Papuan socio-cultural context has not been found. Existing studies only document traditional languages and arts without formulating a knowledge management system rooted in Papuan socio-culture. In fact, the need for a system that is able to combine the power of technology with local wisdom is increasingly urgent, along with the increasing need for indigenous area-based development in Papua. Within this framework, it is important to design a conceptual model of the IKMS cycle that is relevant to the reality of indigenous Papuans. This cycle model is expected to not only be a preservation tool, but also a basis for strengthening customary authority, ecological justice, and community empowerment in the face of social and economic change. Therefore, this research is directed at conceptualizing the Indigenous Knowledge Management System cycle model based on the Papuan socio-cultural context. Thus, the formulation of research questions underlying this study is how to design an Indigenous Knowledge Management System cycle that is relevant to the context of indigenous Papuans, and what are the process elements that need to be included in the IKMS cycle model. By answering this question, it is hoped that it can provide a theoretical contribution to the development of IKMS in the context of indigenous peoples, as well as a practical guide for indigenous peoples, government, academics and development partners in designing the IKMS cycle model.

2 Methodology

This research uses a qualitative method with a literature study approach to examine the Knowledge Management System and Indigenous Knowledge System cycles, and is supported by the emic experience of one researcher as part of the indigenous Papuan community. The literature review focused on mapping previous research relevant to the knowledge management cycle, both from the perspective of modern organizations and indigenous social organizations. Although this study is more likely to be theoretical, it does not merely adopt IKMS cycle models from other regions, but is developed through a process of deep reflection on the emic experiences of researchers who live and are directly involved in the social and cultural dynamics of indigenous Papuans. Therefore, the IKMS cycle developed is not generic, but more contextual and firmly rooted in the social and cultural realities of Papua. The stages to build the conceptual model of the IKMS cycle are carried out in several stages, starting with the collection of scientific work data through publish or perish software using google scholar. The keywords in the search include "knowledge management system" and "indigenous knowledge system" within the period 2000–2025. The data collected consisted of 200 articles for the topic of KMS and 200 for IKS, totaling 400 articles. The data was then selected through a process of inclusion and exclusion, and 22 articles were finally determined. The 22 articles were then mapped (Table 1) and the process elements in the IKMS cycle that reflect the context and socio-cultural reality of Papua were determined. The process elements that have been selected are then depicted in the IKMS cycle (Fig. 1), and the operational definitions of each process element are explained.

3 Results and Discussion

3.1 Mapping of KMS and IKS Processes

Mapping of articles related to knowledge management system and indigenous knowledge system process elements was carried out with inclusion and exclusion criteria. Inclusion criteria include articles published in peer-reviewed journals, relevant topics, full text and reputable journals. Exclusion criteria included duplicate articles, articles that had no substantive relevance to the research topic, and non-empirical publications such as opinion or editorial articles. The screening process involved reviewing the title and abstract, followed by reading the full text. From this process, 22 scientific articles containing the process elements of the knowledge management system and indigenous knowledge were determined, which had the criteria of having at least 4 process elements, as mapped in the table below.

Table 1. KMS and IK Process Mapping

No	References	Summary	Method	findings	KMS & IKS process
Knowledge Management System					
1	[22]	Developing a KMS framework for open and distributed biomedical repositories in Australia	Qualitative: Action research, literature interviews and FGDs	Innovative integrated KMS framework supporting open biomedical knowledge management	Discovery, Creation. Representation, Classification, Storage, Retrieval, Dissemination, Transfer, Translation
2	[23]	Assessing KM practices and technologies in five Kenyan local government counties	Qualitative: Interviews and document analysis of policies and systems.	KM practices are not systematic, technology is limited, integrated policies are needed	Acquisition, Creation, Sharing, Use, Distribution
3	[24]	Analyzing the contribution of KM processes to competitive advantage in private universities in Iraq	Quantitative: Survey	KM processes have a significant effect on the competitive advantage of private universities in Iraq	Identification, Goals Formulating, Generating, Storage, Sharing, Application

(*continued*)

Table 1. (*continued*)

No	References	Summary	Method	findings	KMS & IKS process
4	[25]	Analyzing the role of AI in supporting organizational memory systems based on the seven layers of OMIS	Quantitative: Industry Case Study	AI supports the efficiency of knowledge search, storage, and use in organizations	Source, Acquisition, Formalization, Representation, Processing, Application, Users
5	[26]	AI influences KM processes such as knowledge creation, storage, sharing and application in modern organizations	Qualitative: Literature study and examples of AI application	AI accelerates knowledge creation, storage, and sharing in a strategic and adaptive manner.	Creation, Storing & Retrieving, Sharing, Application
6	[27]	Examines the influence of human behavioral factors on the KMS cycle in universities, especially in creation, transfer, storage and sharing	Quantitative: Survey of academic staff	Motivation and personality factors have a significant effect on the academic KMS cycle	Sharing, Creation, Storing, Transfer
7	[28]	Examining the effect of KM on green product competitive advantage with green process innovation as a mediator in Pakistan	Quantitative: Survey	Knowledge acquisition, storage, and use support green innovation	Acquisition, Creation, Storage & Retrieval, Utilization

(*continued*)

Table 1. (*continued*)

No	References	Summary	Method	findings	KMS & IKS process
8	[29]	Analyzes KM practices in family firms across countries with a focus on unique organizational dynamics and behavioral factors	Qualitative: Literature review and case studies	Family characteristics influence knowledge creation, sharing and retention	Creation, Storage, Sharing, Application
9	[30]	Meta-review of 102 articles on KM exploring barriers and enablers to knowledge transfer and sharing	Qualitative: Metareview	KM is effective in improving organizational learning and adaptation	Acquisition & Generation, Capture, Organization, Storage, Sharing, Application
10	[31]	Analyzing the relationship between KM practices and organizational performance of SMEs with SEM on five key components of KM	Quantitative: Survey	All KM components contribute significantly to improving SME performance	Creation, Acquisition, Sharing, Storage

(*continued*)

Table 1. (*continued*)

No	References	Summary	Method	findings	KMS & IKS process
11	[32]	Modeling knowledge management process enablers using an MCDM approach, with the aim of identifying the main factors that drive successful KM implementation in a systematic and structured manner	Quantitative: ISM and DEMATEL	Knowledge creation and knowledge capture are the main enablers, while knowledge organization and application are dependent, providing practical direction for effective KM implementation in an organizational environment	Protection, Capture, Creation, Dissemination, Storage, Application, Organization, Transfer
Indigenous Knowledge					
1	[33]	Integrating local knowledge with science to address climate change challenges in Southeast Asia	Qualitative: Participatory case study and observation	Identify knowledge that can be integrated or retained	Observation, Documentation, Validation, Categorization
2	[34]	Assessment of KM models in the integration of local and exogenous knowledge for agriculture in Tanzania	Qualitative: Semi-structured interviews	New KM model according to rural social context and local systems	Identification, Acquisition, Development, Sharing, Preservation, Application, Validation
3	[35]	Designing a KM model for the indigenous knowledge system of the Guji Oromo people in Ethiopia	Mixed methods: interview, FGD, questionnaire	Low understanding of IK among young generation without supporting technology system	Identification, Elicitation, Conversion, Modeling, Codification, Evaluation, Filtering

(*continued*)

Table 1. (*continued*)

No	References	Summary	Method	findings	KMS & IKS process
4	[36]	Analysis of indigenous knowledge and impact of ICT in India	Qualitative: Literature study	ICT strengthens the exchange of indigenous knowledge across space and generations	Recognition, Identification, Validation, Documentation, Storage Transfer, Dissemination
5	[37]	An assessment of indigenous knowledge on medical and regional health service regulation	Qualitative: Literature study	Need for legal protection, validation and integration of indigenous knowledge in the health system	Creation, Codification, Utilization, Validation, Protection
6	[38]	Evaluation of IEK's role in conservation and climate and proposed repository in Kenya	Qualitative: Metanalysis	IEK proved useful in conservation, suggested repository and policy support	Identification, Capture, Recording, Structuring, Validation Classification, Publishing, Archives
7	[39]	Identification of indigenous knowledge documentation and dissemination strategies at the University of Zululand	Qualitative: Mixed interpretivist-positivist case study	Limited policies and facilities are a challenge to managing indigenous knowledge	Creation, Sharing, Documenting, Disseminating, Accessing
8	[40]	Evaluating strategies and policies on the protection of indigenous knowledge systems in South Africa	Qualitative: Literature review	The need for a community-based decolonial approach to IKS protection	Creation, Acquisition, Storage, Transfer, Application

(*continued*)

Table 1. (*continued*)

No	References	Summary	Method	findings	KMS & IKS process
9	[41]	Evaluating the role of libraries in the protection of indigenous knowledge in South Africa	Qualitative: Literature study	Libraries need to play an active role in the collection, access and promotion of IK	Collecting, Organizing, Providing access, Identifying, Promoting
10	[42]	Integrating indigenous knowledge into university curricula in Uganda	Qualitative: Interviews, exploratory observation	Need mechanisms for identification, documentation, processing and utilization of IK	Identification, Accessing, Collection, Documentation Organizing, Retrieval, Disseminating, Utilization
11	[43]	Adaptation of the KM model in managing indigenous knowledge in Sub-Saharan African communities	Qualitative: Literature study	The business KM model can be applied in the community with the adaptation of knowledge processes	Identification, Acquisition, Development, Sharing, Utilization, Retention

3.2 Development of Indigenous Knowledge Management System (KMS) Cycle

Based on the literature study in the table above, eight elements of the IKMS cycle were selected in accordance with the needs of the indigenous Papuan community, as described below.

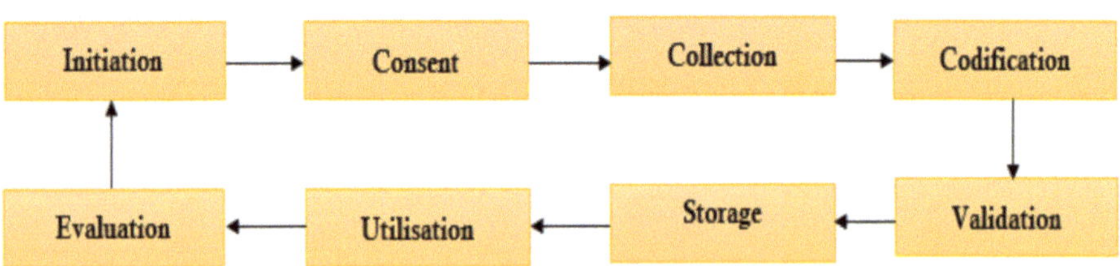

Fig. 1. Indigenous Knowledge Management System Cycle

Initiation. Initiation is one of the early phases in the work system life cycle model [44, 45]. Initiation activities include a planning process consisting of formulating a vision that is in line with the indigenous worldview, setting practical goals, forming teams, allocating resources (technology, funds, manpower, and time) to carry out activities, preparing for FPIC and CARE, and so on. The initiation or planning process is not new; it has existed in indigenous communities for generations. Although not written down, it is carried out systematically and in an organized manner. For example, building traditional houses, gardening, traditional rituals, dance parties, and so on are certainly well planned in terms of the tools needed, the division of roles and responsibilities, consumption, time, and so on. Therefore, it is important to revive and utilize the role of indigenous intellectuals to harmonize indigenous and modern planning values so that they are relevant to the current situation while remaining rooted in local indigenous values.

Consent. Consent is the stage of conscious, voluntary approval by indigenous peoples based on full understanding of an action, decision, or use of their property. Consent is included in the FPIC principle in the United Nations Declaration on the Rights of Indigenous Peoples (UNDRIP) [46], and also one of the important factors in data collection from Melanesian indigenous peoples [47]. The indigenous knowledge possessed by Melanesian indigenous peoples is generally collective in nature and partly contains private (sacred) elements, so there must be collective consent from the knowledge holders. Consent has actually been part of Melanesian tradition for generations. Consent is obtained through a deliberative mechanism that involves important elements of custom. The results are then legalized according to custom through certain rituals or ceremonies. This shows that consent is not only binding between humans, but is also believed to have the blessing of ancestral spirits and cosmological forces that protect the decision. Of course, this is different from modern consent mechanisms that bind between humans, such as through contracts or agreements signed by the parties. Therefore, at this stage, traditional and modern consent mechanisms can be combined to have traditional legitimacy and formal law.

Collection. This stage is carried out after approval has been granted, focusing on the process of collecting, recording, and documenting traditional knowledge. Collection is not limited to oral information, but also includes physical and visual documentation, such as taking plant specimens, recording original names, functions, and parts used, complete with photographic documentation to support the integrity of the data obtained. [48, 49]. In practice, this process must follow technical and ethical guidelines, especially when the knowledge gathered concerns sacred, confidential, or restricted aspects of the community's social system [50].

Codification. Indigenous knowledge that has been collected in the form of text, audio, symbols, objects, videos, images, sketches, and so on through storytelling, interviews, observation, and documentation, is then processed and transformed into an explicit, structured, and valuable form. This process aims to convert tacit knowledge that is contextual and passed down orally and in the form of objects or symbols into a form that can be stored, accessed, and reused, such as text, images, videos, audio, and digital maps [51, 52]. Codification ensures that the knowledge that has been gathered is not only documented, but also can be used across generations in a sustainable manner without neglecting inherent cultural values [53]. Codification is carried out with consideration for narrative structure, symbolism, and regional language, a process that is directly supervised by knowledge holders to maintain the integrity and validity of the content [54].

Validation. The validation stage is a systematic process to ensure that the codified knowledge is accurate, relevant, consistent, and valid according to the indigenous community. Validation serves as an important bridge between codification and storage, as it not only confirms the validity of the knowledge content, but also determines its accessibility status, whether it is private as an internal community archive or can be shared publicly with certain restrictions [55]. This process includes verification of the structure and coherence of knowledge, as well as evaluation of its usefulness, cultural sensitivity, and the level of public trust in that knowledge [56]. Validation is carried out through participatory dialogue with knowledge holders and the younger generation, in order to maintain the integrity and authenticity of information, while aligning it with the collective principles of indigenous peoples [54]. At this stage, the storage method is also determined, whether in digital or conventional form, which is appropriate and accepted by the local indigenous community. At this stage, collective approval of the codification results is also carried out, which can be done through customary approval mechanisms and ratified through rituals or traditional ceremonies to obtain the blessing of ancestral spirits and cosmological forces that protect the decision. Thus, validation is not only technical but also ethical in the context of the local indigenous community.

Storage. The storage stage is an important process in maintaining the continuity and sustainability of indigenous knowledge that has undergone the customary validation stage. At this stage, knowledge that has been confirmed to be valid and culturally sensitive will be systematically archived, either through digital media or conventional means such as manuscripts, images, recordings, and physical artifacts, according to the preferences of the indigenous community [57]. Knowledge storage acts as the collective memory of a community, preventing the loss of important information due to generational change or the loss of key knowledge holders [58]. The existence of a digital repository enables effective and efficient access and utilization for the purpose of passing on knowledge to the younger generation and also for development based on traditional values.

Utilisation. The utilization stage is the process of applying indigenous knowledge that has been stored to support concrete actions that benefit indigenous communities. Previously documented knowledge is insufficient if it is not applied in the context of daily life, such as collective decision-making, culture-based education, natural resource management, disaster mitigation, conflict management, development planning and so on [50]. Therefore, utilization in IKMS emphasizes the need to integrate indigenous knowledge into sustainable development policies. This process also requires user absorptive capacity, namely the ability of individuals or indigenous peoples to understand, adopt, and apply knowledge effectively [37, 59]. In addition, the utilization of indigenous knowledge aims to create social and ecological excellence through innovations that are in line with traditional values, as well as strengthening cultural resilience amid social and environmental changes [28]. Thus, the utilization stage is important to ensure that the indigenous knowledge that has been collected, codified, validated, and stored has a direct impact on the sustainability and independence of indigenous peoples.

Evaluation. Evaluation is an important stage in management science [60–62]. This stage is crucial for measuring the effectiveness of planning, collecting, codifying, verifying, storing, and utilizing traditional knowledge, as well as identifying weaknesses, errors, obstacles, or opportunities in knowledge management. Based on the evaluation results, tactical and strategic steps need to be taken to make continuous improvements. After this stage, the cycle will repeat to ensure technical and ethical system updates.

4 Conclusions

This research supports the idea that the digitization of indigenous knowledge should be designed on the basis of the social, cultural, ecological and spiritual structures of indigenous peoples [63]. Based on this understanding, this research develops an IKMS cycle that is contextualized with the reality of indigenous Papuans. This cycle is expected to be a reference in managing IK systematically and sustainably. More than just a technical guide, this cycle is positioned as a mechanism to integrate information technology with local knowledge to strengthen the cultural identity and autonomy of indigenous peoples. Based on the literature review and emic experience of the researchers, seven elements of the IKMS cycle process were formulated, namely Initiation, Consent, Collection, Codification, Validation, Storage, Utilization and Evaluation. These eight elements are the result of integration between the knowledge management system process and indigenous knowledge. Thus, the IKMS cycle designed not only adopts the principles of knowledge management in modern organizations, but is also rooted in the cultural values of indigenous Papuans, thus demanding ethical sensitivity and contextualization. Therefore, every stage in the IKMS cycle must apply the principle of Free, Prior and Informed Consent (FPIC) to ensure that indigenous peoples become active subjects, not passive objects of others [64]. This strengthens social legitimacy and ensures a fair and responsible knowledge management system. Thus, the IKMS is not a documentation tool, but a socio-cultural framework to strengthen the independence and autonomy of indigenous Papuans.

References

1. Onyancha, O.B.: "Indigenous knowledge, traditional knowledge and local knowledge: what is the difference? An informetrics perspective. Glob. Knowl. Memory Commun. (2022). https://doi.org/10.1108/GKMC-01-2022-0011
2. Sokoy, F.: Economic development of indigenous papuans: a critical analysis to build independent and dignified traditional villages. J. Soc. Sci. **3**(6), 1233–1246 (2022). https://doi.org/10.46799/jss.v3i6.437
3. Ademowo, A.J., Nuhu, A.A.: Indigenous knowledge and conflict management in Africa: a study of proverb use in conflict management among Hausas of Northern Nigeria. Int. J. History (2017). [Online]. Available: https://www.academia.edu/download/96925892/e9c78bc8e45f4b6f2154415d2099d1b2f389.pdf
4. Olu, O.I., Kanu, I.A.: Mitigation ofAfrican Indigenous Knowledge. J. Afr. Stud. Sustain. (2021). [Online]. Available: https://www.acjol.org/index.php/jassd/article/view/1498
5. Verma, P.: Integrating indigenous knowledge and western science into forestry, natural resources, and environmental programs. J. For. **114**(6), 648–655 (2016). https://doi.org/10.5849/jof.15-090

6. Bello-Bravo, J., Lutomia, A.N.: Sustainable development or developmental sustainability: two cases of indigenous knowledge and practices for sustainable sourcing for wood-based design-solutions. Elsevier B.V. (2022). https://doi.org/10.1016/j.tfp.2022.100253
7. Doyo, M.C., Kumar, R.: Empowering forest conservation: harnessing indigenous knowledge within Southern Ethiopia's forest-dependent communities. J. Manage. Serv. (2024). [Online]. Available: http://jmss.a2zjournals.com/index.php/mss/article/view/59
8. Ullah, S.: Indigenous knowledge, climate change and transformations of Gwadar fishing community. Int. J. Clim. Chang. Strateg. Manag. (2022). https://doi.org/10.1108/IJCCSM-06-2022-0069
9. Quigley, C.: Globalization and science education: the implications for indigenous knowledge systems. Int. Educ. Stud. (2009). [Online]. Available: https://eric.ed.gov/?id=EJ1065538
10. Suharyanto and Wiflihani: Preserving local culture in the era of globalization: balancing modernity and cultural identity. Path Sci. **10**(3), 5001–5005 (2024). https://doi.org/10.22178/pos.102-16
11. Warami, H.: Pemetaan Wilayah Adat: Identifikasi dan Registrasi Wilayah Adat Suku di Papua Barat. Manokwari (2019). Accessed: 18 Jan 2026. [Online]. Available: https://www.researchgate.net/profile/Hugo-Warami/publication/348002265_PEMETAAN_WILAYAH_ADAT/links/5fec7e7f92851c13fed4156a/PEMETAAN-WILAYAH-ADAT.pdf
12. Warwer, F., Pontoan, D.R.: Ondofolo's leadership in Christian perspective: a study based on Papuan-Indonesian local wisdom. Verbum et Ecclesia **44**(1) (2023). https://doi.org/10.4102/ve.v44i1.2773
13. Pamungkas, C., Oktafiani, I., Imbhiri, L.: Makna Pembangunan Bagi Orang Asli Papua: Studi Terhadap Marginalisasi dan Depopulasi di Tanah Papua. Jurnal Ilmiah Hubungan Internasional 16–38 (2022). https://doi.org/10.26593/jihi.v0i00.5970.16-38
14. Hidayatulloh, H., Erdős, É., Szabó, M.: The intricate justice of poverty: a case of the land of Gold in Papua Indonesia. J. Indonesian Legal Stud. **7**(2), 557–584 (2022). https://doi.org/10.15294/jils.v7i2.58030
15. Jasper, P., Jochem, W.C., Lambert-Porter, E., Naeem, U., Utazi, C.E.: Mapping the prevalence of severe acute malnutrition in Papua, Indonesia by using geostatistical models. BMC Nutr. **8**(1), 13 (2022). https://doi.org/10.1186/s40795-022-00504-z
16. Lefaan, Y., Mollet, J.A., Innah, H.: The sustainable development in Papua based on customary territories and human development paradigm. Int. J. Soc. Adm. Sci. **6**(2), 86–96 (2021). https://doi.org/10.18488/journal.136.2021.62.86.96
17. Ngabalin, Adi, Tampake: Pemalangan Tanah Sebagai Bentuk Resistensi Masyarakat Papua Barat Terhadap Pembangunan. Indonesian J. Relig. Soc. **6**(2), 66–77 (2024). Accessed: 18 Jan 2026. [Online]. Available: https://journal.lasigo.org/index.php/IJRS/article/view/423
18. Ghimire, P.: Digitalization of Indigenous knowledge in Nepal–Review article. Acta Informatica Malays. **5**(2), 42–47 (2021). https://doi.org/10.26480/aim.02.2021.42.47
19. Mwanza, A.J.: Can digital technology be used for addressing digital literacy in the dissemination of Indigenous Knowledge in rural areas?. South Afr. J. Libr. Inf. (2022). [Online]. Available: https://sajlis.journals.ac.za/pub/article/view/2137
20. Sitokdana, M.N.N., Tanone, R., Tanaem, P.F.: Digitalization of the local language dictionary of Pegunungan Bintang. Procedia Comput. Sci. **161**, 49–56 (2019). https://doi.org/10.1016/j.procs.2019.11.098
21. Sibarani, Munthe, Irviantina: Pemanfaatan Teknologi Informasi Guna Meningkatkan Pemasaran Produk Masyarakat Di Kampung Nelayan Sebrang Medan Belawan. Kumawula: Jurnal Pengabdian Kepada Masyarakat **5**(2), 309 (2022). https://doi.org/10.24198/kumawula.v5i2.37100
22. Kruesi, L., Burstein, F., Tanner, K.: A knowledge management system framework for an open biomedical repository: communities, collaboration and corroboration. J. Knowl. Manag. **24**(10), 2553–2572 (2020). https://doi.org/10.1108/JKM-05-2020-0370

23. Ndiege, J.R.A., Wamuyu, P.K.: Knowledge management practices and systems in county governments in developing countries. VINE J. Inf. Knowl. Manage. Syst. ahead-of-print (2019). https://doi.org/10.1108/VJIKMS-01-2019-0014
24. Mahdi, O.R., Nassar, I.A., Almsafir, M.K.: Knowledge management processes and sustainable competitive advantage: an empirical examination in private universities. J. Bus. Res. **94**, 320–334 (2019). https://doi.org/10.1016/j.jbusres.2018.02.013
25. Alstete, J.W., Meyer, J.P.: Intelligent agent-assisted organizational memory in knowledge management systems. VINE J. Inf. Knowl. Manage. Syst. **50**(4), 615–630 (2020). https://doi.org/10.1108/VJIKMS-05-2019-0063
26. Jarrahi, M.H., Askay, D., Eshraghi, A., Smith, P.: Artificial intelligence and knowledge management: a partnership between human and AI. Bus. Horiz. **66**(1), 87–99 (2023). https://doi.org/10.1016/j.bushor.2022.03.002
27. Rashid, A.S., Tout, K., Yakan, A.: The critical human behavior factors and their impact on knowledge management system–cycles. Bus. Process. Manag. J. **27**(6), 1677–1702 (2021). https://doi.org/10.1108/BPMJ-11-2020-0508
28. Rehman, S., Asif, T., Nasir, A., Ullah, A., Noor, N.: Harnessing knowledge management for green product competitive advantage with green process innovation as a mediator. J. Open Innovation: Technol. Market Complex. **10**(2), 100274 (2024). https://doi.org/10.1016/j.joitmc.2024.100274
29. Arzubiaga, U., De Massis, A., Kammerlander, N., Hoy, F.: Guest editorial. J. Knowl. Manag. **26**(2), 269–290 (2022). https://doi.org/10.1108/JKM-03-2022-924
30. Fierro, R., Benalil, G.: Benefits, process and challenges of knowledge management. J. Enterp. Bus. Intell. 083–094 (2024). https://doi.org/10.53759/5181/JEBI202404009
31. Gholami, A., Shirkouhi, N.: Investigating the influence of knowledge management practices on organizational performance: an empirical study. Acta Polytechnica Hungarica **10**(2), 205–2016 (2013). Accessed: 18 Jan 2026. [Online]. Available: https://acta.uni-obuda.hu/Gholami_Asli_Nazari-Shirkouhi_Noruzy_40.pdf
32. Agrawal, N.: Modeling enablers of knowledge management process using multi criteria decision making approach. VINE J. Inf. Knowl. Manage. Syst. **51**(3), 389–417 (2021). https://doi.org/10.1108/VJIKMS-08-2019-0122
33. Hiwasaki, L., Syamsidik, E.L., Shaw, R.: Process for integrating local and indigenous knowledge with science for hydro-meteorological disaster risk reduction and climate change adaptation in coastal and small island communities. Int. J. Disaster Risk Reduction **10**, 15–27 (2014). https://doi.org/10.1016/j.ijdrr.2014.07.007
34. Tandi Lwoga, E.: Knowledge management approaches in managing agricultural indigenous and exogenous knowledge in Tanzania. J. Documentation **67**(3), 407–430 (2011). https://doi.org/10.1108/00220411111124523
35. Sorsa, Haro, Kamaru: Designing indigenous knowledge management model for gadaa system: rule-based knowledge representation approach. J. Indigenous Knowl. Dev. Stud. **2**(2), 1–19 (2020). Accessed: 18 Jan 2026. [Online]. Available: https://ejol.aau.edu.et/index.php/JIKDS/article/download/3776/2763/6136
36. Sharma, A.K.: Indigenous knowledge communication in the 21st Century. Int. J. Digital Libr. Serv. **4**(2), 128–135 (2014). Accessed: 18 Jan 2026. [Online]. Available: http://www.ijodls.in/uploads/3/6/0/3/3603729/ajay_kum_sharma_128-135.pdf
37. Ouma, C.F.O.: Issues to consider when applying indigenous medical knowledge to provide healthcare by county governments in Kenya. Regional J. Inf. Knowl. Manage. **7**(2), 14–37 (2022). Accessed: 18 Jan 2026. [Online]. Available: https://rjikm.org/index.php/rjikm/article/view/59

38. Guto, R.: A meta-Analytical review of the role of indigenous knowledge on environmental conservation and climate change in Kenya. J. Inf. Knowl. Manage. **5**(2), 65–84 (2020). Accessed: 18 Jan 2026. [Online]. Available: https://www.ajol.info/index.php/rjikm/article/view/243877
39. Buthelezi, S.C., Ocholla, D., Dlamini, P.: Strategies for documenting and disseminating indigenous knowledge at a South African university. SA J. Inf. Manage. **26**(1) (2024). https://doi.org/10.4102/sajim.v26i1.1648
40. Masenya, T.M.: Decolonization of indigenous knowledge systems in South Africa. Int. J. Knowl. Manage. **18**(1), 1–22 (2022). https://doi.org/10.4018/IJKM.310005
41. Lor, P.: Storehouses Of Knowledge? The Role Of Libraries In Preserving And Promoting Indigenous Knowledge. Afr. J. Indigenous Knowl. Syst. **3**(1), 45–56 (2004). Accessed: 18 Jan 2026. [Online]. Available: https://hdl.handle.net/10520/EJC61434
42. Magara, E.: Integration of indigenous knowledge management into the university curriculum: a case For Makerere University. Afr. J. Indigenous Knowl. Syst. **14**(1), 25–41 (2015). Accessed: 18 Jan 2026. [Online]. Available: https://hdl.handle.net/10520/EJC174406
43. Ngulube, Lwoga: Knowledge management models and their utility to the effective management and integration of indigenous knowledge with other knowledge systems. Afr. J. Indigenous Knowl. Syst. **6**(2), 117–131 (2007). Accessed: 18 Jan 2026. [Online]. Available: https://www.researchgate.net/publication/259801712_Knowledge_Management_Models_And_Their_Utility_To_The_Effective_Management_And_Integration_Of_Indigenous_Knowledge_With_Other_Knowledge_Systems
44. Alter, S.: The work system method for understanding information systems and information systems research. Commun. Assoc. Inf. Syst. **9** (2002). https://doi.org/10.17705/1CAIS.00906
45. Alter, S.: Work system theory: overview of core concepts, extensions, and challenges for the future. J. Assoc. Inf. Syst. **14**(2), 72–121 (2013). Accessed: 18 Jan 2026. [Online]. Available: https://aisel.aisnet.org/jais/vol14/iss2/1/
46. Raftopoulos, M., Short, D.: Implementing free prior and informed consent: the United Nations Declaration on the Rights of Indigenous Peoples (2007), the challenges of REDD+ and the case for the precautionary principle. Int. J. Hum. Rights **23**(1–2), 87–103 (2019). https://doi.org/10.1080/13642987.2019.1579990
47. Vallance, R.J.: A Melanesian research methodology. In: AARE/APERA International Conference, Sydney: Australian Catholic University, Strathfield., pp. 2–16 (2012). Accessed: 18 Jan 2026. [Online]. Available: https://www.aare.edu.au/data/publications/2012/Vallance12.pdf
48. Mahwasane, S.T., Middleton, L., Boaduo, N.: An ethnobotanical survey of indigenous knowledge on medicinal plants used by the traditional healers of the Lwamondo area, Limpopo province, South Africa. S. Afr. J. Bot. **88**, 69–75 (2013). https://doi.org/10.1016/j.sajb.2013.05.004
49. Khakurel, D., Uprety, Y., Łuczaj, Ł, Rajbhandary, S.: Foods from the wild: local knowledge, use pattern and distribution in Western Nepal. PLoS ONE **16**(10), e0258905 (2021). https://doi.org/10.1371/journal.pone.0258905
50. Magara, E.: Integration of indigenous knowledge management into the university curriculum: a case for Makerere University. Indilinga Afr. J. Indigenous Knowl. (2015). https://doi.org/10.10520/EJC174406
51. Ngulube, P.: Managing and preserving indigenous knowledge in the knowledge management era: challenges and opportunities for information professionals. Inf. Dev. (2002). https://doi.org/10.1177/026666602400842486
52. Masango, C.A.: Indigenous knowledge codification of African traditional medicine: inhibited by status quo based on secrecy? Inf. Dev. **36**(3), 327–338 (2020). https://doi.org/10.1177/0266666919853007

53. Falola, T.: Codification, documentation, and transmission of knowledge in African traditional religion. In: The Palgrave Handbook of African Traditional Religion, Cham: Springer International Publishing, pp. 497–514 (2022). https://doi.org/10.1007/978-3-030-89500-6_38
54. Echajari, L., Thomas, C.: Learning from complex and heterogeneous experiences: the role of knowledge codification. J. Knowl. Manag. **19**(5), 968–986 (2015). https://doi.org/10.1108/JKM-02-2015-0048
55. Owoc, M.L., Ochmanska, M., Gladysz, T.: On principles of knowledge validation. In: Validation and Verification of Knowledge Based Systems, Boston, MA: Springer US, pp. 25–35 (1999). https://doi.org/10.1007/978-1-4757-6916-6_2
56. Vicat, C.: Knowledge validation in the building of a knowledge-based system. Expert Syst. Appl. **8**(3), 391–397 (1995). https://doi.org/10.1016/0957-4174(94)E0030-X
57. Durst, S., Edvardsson, I.R., Foli, S.: Knowledge management in SMEs: a follow-up literature review. J. Knowl. Manag. **27**(11), 25–58 (2023). https://doi.org/10.1108/JKM-04-2022-0325
58. Zaim, H., Muhammed, S., Tarim, M.: Relationship between knowledge management processes and performance: critical role of knowledge utilization in organizations. Knowl. Manag. Res. Pract. **17**(1), 24–38 (2019). https://doi.org/10.1080/14778238.2018.1538669
59. Ngulube, P., Lwoga, E.: Knowledge management models and their utility to the effective management and integration of indigenous knowledge with other knowledge systems. Afr. J. Indigenous Knowl. (2007). https://doi.org/10.10520/EJC61532
60. Sadeghifar, J., Jafari, M., Tofighi, S., Ravaghi, H., Maleki, M.R.: Strategic planning, implementation, and evaluation processes in hospital systems: a survey from Iran. Glob. J. Health Sci. **7**(2) (2014). https://doi.org/10.5539/gjhs.v7n2p56
61. Kerzner, H.: Project Management: A Systems Approach to Planning, Scheduling, and Controlling, 10th edn. John Wiley & Sons Inc, Hoboken, New Jersey (2009)
62. Kabeyi, M.J.B.: Organizational strategic planning, implementation and evaluation with analysis of challenges and benefits for profit and nonprofit organizations. Int. J. Appl. Res. **5**(6), 27–32 (2019). https://doi.org/10.22271/allresearch.2019.v5.i6a.5870
63. Jimenez, A., Hoefsloot, F.I., Miranda Sara, L.: Towards decolonial IS: Insights from applying pluriverse and conviviality to analyse a co-production intervention in Peru. Inf. Syst. J. **35**(3), 907–932 (2025). https://doi.org/10.1111/isj.12565
64. Hiwasaki, L., Luna, E., Shaw, R.: Process for integrating local and indigenous knowledge with science for hydro-meteorological disaster risk reduction and climate change adaptation in. Int. J. Disaster Risk (2014). [Online]. Available: https://www.sciencedirect.com/science/article/pii/S2212420914000612

Why Users Complain: Identifying Service Quality Gaps in Nepal's Leading e-wallets

Kapil Gurung(✉)

Seoul Christian University, Seoul, South Korea
kapilgurung@scu.ac.kr

Abstract. This study examines Nepal's three leading electronic wallets, eSewa, Khalti, and IMEPay, to assess consumer sentiment toward service quality by analyzing 3,000 user-generated reviews from the Google Play Store. Applying the e-SERVQUAL framework and sentiment analysis, the research identifies key strengths and service gaps across four dimensions: efficiency, fulfillment, system availability, and privacy, particularly for eSewa and Khalti. IMEPay demonstrates comparatively more positive sentiment, largely driven by its user-friendly interface, although concerns remain regarding transaction reliability and data privacy. Overall, negative sentiment dominates across all platforms, emphasizing the need for improved technical stability, transparent policies, and stronger data protection measures. The study highlights critical implications for service providers and regulators, emphasizing the importance of enhancing reliability, strengthening privacy safeguards and optimizing customer support to foster long-term trust and user satisfaction in Nepal's digital payment ecosystem.

Keywords: UGC · e-wallet · e-service quality · sentiment analysis

1 Introduction

The rapid growth of digital technologies has transformed financial transactions worldwide, with payment systems becoming a central force in modern economies [1]. Since the early 2000s, pioneers such as PayPal and Alipay have laid the foundation for seamless online payments, enabling the rise of integrated digital ecosystems that integrate e-commerce, social networking, and financial services. A digital wallet, often called an e-wallet, is an electronic application that allows users to store funds, link bank accounts or cards, and make secure transactions directly through smartphones or other digital devices [2]. These systems serve as convenient alternatives to cash, offering speed, accessibility, and financial inclusion. This evolution has improved accessibility and financial inclusion, driven by smartphone penetration and platforms like Apple Pay and Google Pay, and regional players adapting to local needs. The digital payment market is projected to exceed 20 trillion dollars by 2025 globally, reflecting the shift from cash-based economies to digital-first systems [3]. This global trend has also reflected the rise of digital payment platforms in developing economies such as Nepal. Among many e-wallets, eSewa, Khalti and IMEPay are the top three that are integral to the country's financial landscape [4].

H.-S. Ma and Y.-W. Chan (Eds.): IC 2026, LNEE 1646, pp. 110–118, 2027.
https://doi.org/10.1007/978-981-92-0779-4_12

The adoption of mobile wallets in Nepal has transformed the payment ecosystem, providing a viable alternative to cash-based transactions and promoting financial inclusion across diverse demographics [5]. The growing smartphone usage rate (72.5%) and expanding high-speed internet connectivity has enabled these services to deliver unmatched convenience, speed, and accessibility in daily life [6]. However, as digital payment systems become more embedded in Nepal's financial ecosystem, the quality of these services remains a critical factor influencing user satisfaction and long-term adoption. With more than a dozen digital payment systems available, competition in the online payment sector is intense. This environment compels major providers such as eSewa, Khalti, and IMEPay to adopt proactive strategies to retain users, particularly by enhancing service quality. Strong service quality fosters trust and cultivate long-term relationships, while poor service drives customers toward competitors, reducing market share and profitability [7]. In Nepal, where digital payment infrastructure is still evolving, users have reported mixed experiences, including concerns over transaction security, data privacy, and inadequate grievance-handling mechanisms [8]. Since service quality affects customer satisfaction, achieving excellence is vital. High satisfaction levels not only strengthen trust and loyalty but also contribute positively to financial performance through increased sales and profitability. To support continuous improvement, the electronic service quality (e-SERVQUAL) framework offers a reliable means of assessing digital service quality. According to Parasuraman, Zeithaml and Malhotra [9], it encompasses four key dimensions: efficiency, fulfillment, system availability, and privacy. For Nepal's digital wallet providers to remain competitive, excelling in these dimensions is essential to ensure a seamless and secure user experience that meets modern expectations.

This study aims to evaluate customer sentiment toward three major e-wallets (eSewa, Khalti, and IMEPay) through the lens of the e-SERVQUAL framework. While most companies rely on traditional methods such as surveys and statistical analyses, these approaches often fail to capture the nuanced emotions and spontaneous opinions expressed by users in real-time settings [10]. Recent studies in Nepal also demonstrate that user-generated content on digital platforms reflects authentic public sentiment more reliably than traditional surveys, especially when computational methods such as sentiment analysis are employed [11, 12]. To address this gap, this study applies sentiment analysis, a natural language processing technique that systematically extracts and interprets positive, negative, or neutral sentiments from unstructured textual data. As illustrated in Fig. 1., negative sentiments dominate customer perceptions of these e-wallets, reflecting dissatisfaction with current offerings. Therefore, it is essential to understand the specific elements of e-service quality driving this dissatisfaction.

Drawing on user-generated content (UGC) from the Google Play Store, this research identifies user sentiments, pinpoints service weaknesses, and provides actionable recommendations for improving Nepali's digital payment ecosystem. The findings offer valuable implications for service providers and policymakers, highlighting strengths to leverage and weaknesses to address to achieve enhanced user experience.

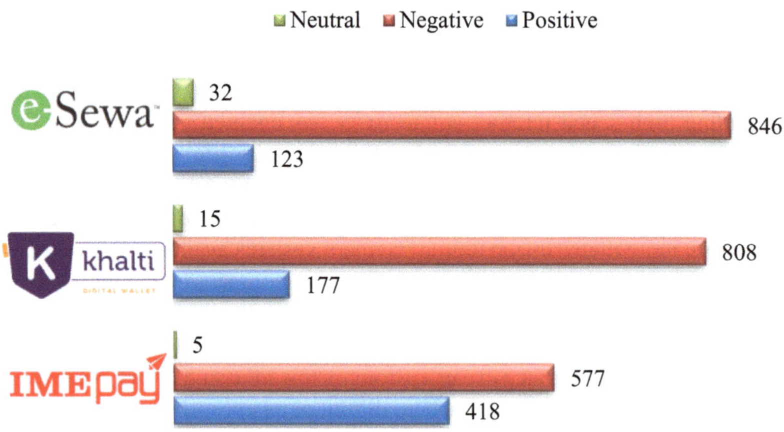

Fig. 1. Comparative User sentiments for e-wallets: eSewa, Khalti and IMEPay

2 Literature Review

2.1 E-Service Quality and Digital Payments

E-service quality refers to the overall effectiveness, efficiency, and user satisfaction derived from online service delivery in virtual environments [7]. It captures a customer's entire experience, from navigating a platform to completing transactions. The e-SERVQUAL model, originally developed by Parasuraman, Zeithaml and Malhotra [9], is a widely recognized framework for evaluating electronic service quality. It has evolved into one of the most robust frameworks for analyzing online service quality, adapting to the complexities of modern virtual interactions. This model identifies four core dimensions to measure e-service quality: efficiency, fulfillment, system availability, and privacy. In the context of fintech applications such as e-wallets, each dimension is defined as follows:

a. Efficiency reflects the speed of login, KYC (know your customer) processing, and transaction execution, critical for users who rely on mobile payments for daily activities.
b. Fulfillment corresponds to the accuracy and reliability of fund transfers, bill payments, and cashback settlements.
c. System availability is tied to the platform's technical robustness, including uptime stability, server responsiveness, and QR code functionality.
d. Privacy reflects the platform's handling of sensitive user information and such as citizenship details, biometric data and GPS access required for KYC and compliance.

These mappings ensure that the e-SERVQUAL framework captures the fintech-specific attributes most relevant to user trust and adoption.

2.2 Sentiment Analysis and Digital Payments

Sentiment analysis, or opinion mining, uses natural language processing and machine learning to extract and classify emotions from text [13]. It categorizes sentiments as positive, negative, or neutral, providing businesses with insights into customer perceptions. In e-service contexts, sentiment analysis quantifies satisfaction across specific dimensions, offering a cost-effective alternative to surveys [14]. Its application in digital financial services has grown, with studies like Adiningtyas and Auliani [15] demonstrating its utility in assessing mobile banking quality. Furthermore, Gurung [16] highlights that modern sentiment analysis techniques, using deep learning-cross-platform analytics, comments, and emoji-based affective signals, offer richer insights than traditional self-reported data, thereby reinforcing the value of computational methods in understanding user experiences. The study adapts this approach to digital payment systems, analyzing UGC to uncover service quality gaps.

3 Methodology

The research follows a four-step workflow: Data collection and preprocessing, e-SERVQUAL dimension classification, and sentiment classification.

3.1 Data Collection and Preprocessing

A total of 3,000 user reviews (1,000 each for eSewa, Khalti, and IMEPay) were collected in compliance with the Google Play Store's Terms of Service, spanning the period from April 4, 2023, to December 31, 2024. All the reviews were de-identified before storage, and analysis was reported in aggregate form. These e-wallets were selected based on their popularity in Nepal's digital payment market, where eSewa is the oldest and most downloaded app with the most reviews compared to the other two (see Table 1). The collected reviews were cleaned by converting the text to lowercase, removing non-alphabetic characters, tokenizing into words, eliminating stop words, and stemming to their root forms. This structured the unstructured UGC for analysis.

Table 1. Overview of three major e-wallets in Nepal

	Release date	App downloads	Rating/Reviews
eSewa	Dec 31, 2012	10M +	3.4/76k
Khalti	May 29, 2016	5M +	3.5 /25k
IMEPay	May 18, 2017	1M +	4.1/24k

3.2 E-SERVQUAL Dimension Classification

A total of 500 review samples were manually annotated for sentiment and the four e-SERVQUAL by two independent experts. Inter-rater reliability was high (Cohen's Kappa = 0.82), and disagreements were resolved before finalizing the training dataset. A Naïve Bayes classifier was then implemented in Google Collaboratory, with the data split into 70% training and 30% testing sets. The model automatically categorized reviews into four e-SERVQUAL dimensions and three sentiment types. The sentiment classifier achieved 89.4% accuracy and an F1-score of 0.87(positive),0.90(negative), and 0.81(neutral). The e-SERVIQUAL classifier reached 86.2% accuracy with a macro-averaged F1-score of 0.84. These metrics confirm that the classifications used in the analysis are robust and trustworthy.

3.3 Sentiment Classification

The same classifier identified sentiments as positive, negative, or neutral within each dimension. Table 2 shows a four-dimensional e-SERVQUAL classification anonymized excerpt along with positive and negative sentiments. Following Adiningtyas and Auliani [15], this study modified its methodology for application in digital payment contexts, thereby ensuring accurate sentiment classification.

Table 2. Classification of E-Service quality dimension review excerpt with sentiment

Reviews Excerpt (Anonymized)	E-SERVQUAL Dimension	Sentiment
Khalti is the best digital wallet, offering a fast and smooth experience for all my transactions	Efficiency	Positive
I transferred money from Khalti to my bank, but it hasn't arrived after days, leaving me frustrated with no clear refund timeline	Fulfillment	Negative
This app runs smoothly compared to other apps, which often crash. It is making us reliable for daily use	System Availability	Positive
Forcing me to upload a selfie with my citizenship for KYC, visible to all, is a serious privacy violation	Privacy	Negative

4 Results and Discussion

4.1 E-SERVQUAL Dimension Classification Result

The retrieved data is categorized according to the four service quality dimensions. This step utilizes four e-SERVQUAL dimensions to evaluate the service quality of e-wallets. Figure 2, presents the results of the text classification analysis. Among the dimensions, efficiency emerges as the strongest aspect across all three platforms, with IMEPay (487) slightly overperforming Khalti (431) and eSewa (412). Fulfillment ranks second in IME-Pay (258) but third for eSewa (196) and Khalti (219). System availability, although lower than fulfillment in IMEPay (207) scored higher in eSewa (297) and Khalti (280). On the other hand, privacy is consistently rated the lowest across all three platforms, with minor variations.

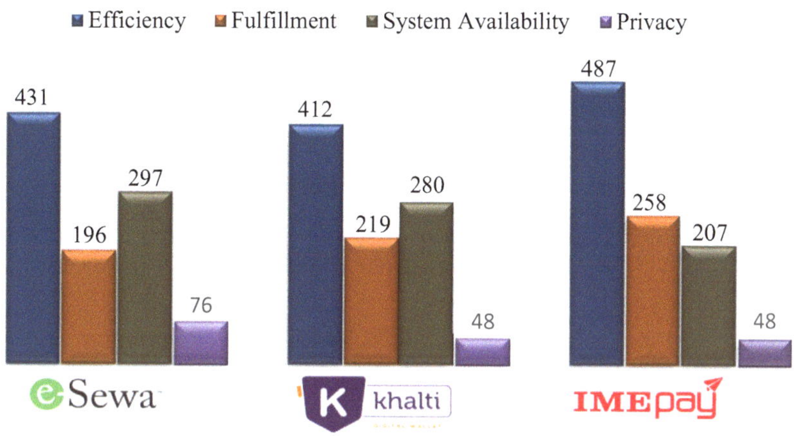

Fig. 2. e-SERVQUAL dimensions for e-wallets: eSewa, Khalti and IMEPay

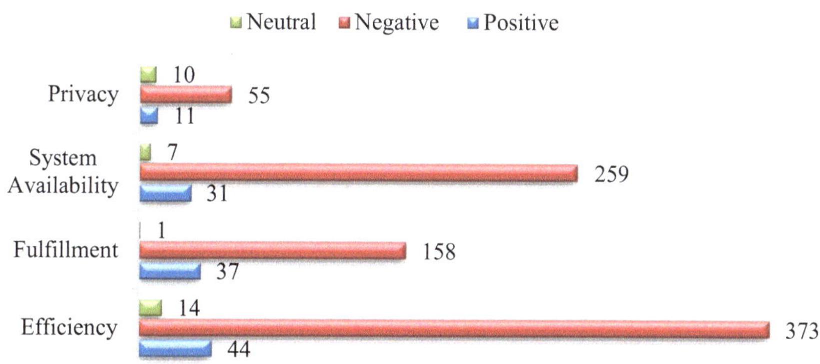

Fig. 3. Sentiment analysis of e-SERVQUAL dimensions for eSewa

4.2 Sentiment Analysis of e-SERVQUAL Dimensions for e-wallets

For eSewa, negative sentiment dominated across all e-service quality dimensions. According to Fig. 3., Efficiency (431 reviews, 43.1%) was heavily negative (373 reviews, 86.5%) due to KYC verification issues, login problems, and MPIN errors, with positive (44, 10.2%) noting ease of use when functional. Fulfillment (196, 19.6%) saw negative feedback (158, 80.6%) for stuck funds, failed transactions, and delayed payments, with positive reviews (37, 18.9%) praising successful transactions. System Availability (297, 29.7%) was overwhelmingly negative (259, 87.2%) due to frequent crashes, server issues, and app inaccessibility, with few positive reviews (31, 10.4%) citing reliability. Privacy (76, 7.6%) was largely negative (55, 72.4%) due to data security concerns, forced location access, and KYC privacy issues, with positive reviews (11, 14.5%) highlighting trust. Overall, eSewa's sentiment is 12.3% positive, 84.5% negative, and 3.2% neutral, reflecting greater dissatisfaction compared to Khalti and IMEPay.

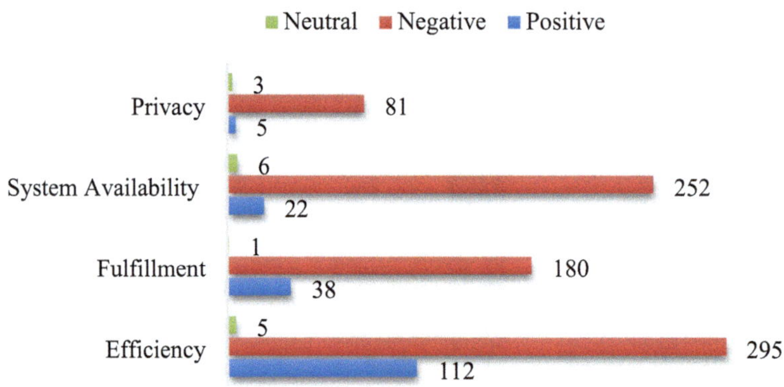

Fig. 4. Sentiment analysis of e-SERVQUAL dimensions for Khalti

For Khalti, Efficiency (412 reviews, 41.2%) is mostly negative (295, 71.6%) due to KYC and login issues, with some positive feedback (112, 27.2%). Fulfillment (219 reviews, 21.9%) is largely negative (180, 82.2%) for delayed or failed transactions, with few positive reviews (38, 17.4%). System Availability (280, 28.0%) is heavily negative (252, 90.0%) due to crashes and server issues, with minimal positive reviews (22, 7.9%). Privacy (89, 8.9%) is predominantly negative (81, 91.0%) for forced GPS and KYC privacy concerns, with rare positive feedback (5, 5.6%). Khalti's sentiment is 17.7% positive, 80.8% negative, and 1.5% neutral (See Fig. 4).

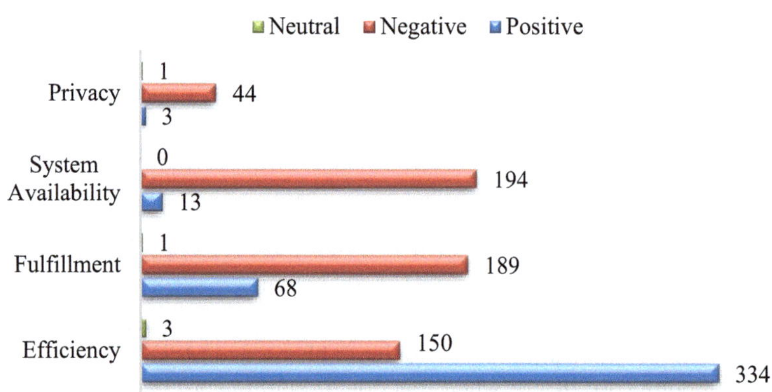

Fig. 5. Sentiment analysis of e-SERVQUAL dimensions for IMEPay

IMEPay demonstrates a mixed performance across e-service quality dimensions while receiving relatively more positive sentiment (418 reviews, 41.8%) compared to other platforms (see Fig. 5). Efficiency, the most frequently mentioned dimension (487, 48.7%), is the only category with predominantly positive sentiment (334, 68.6%), driven by praise for its user-friendly interface and ease of use. Fulfillment (258, 25.8%) is largely negative (189, 73.3%) due to transaction failures, misleading cashback offers, and unexpected charges. System Availability (207, 20.7%) is also overwhelmingly negative (194,

93.7%), reflecting frequent app crashes, QR code scanning issues, and slow loading times. Lastly, privacy (48, 4.8) is also predominantly negative (44, 91.7%) due to concerns about unclear security protocols, KYC privacy issues, and challenges in deleting accounts. Overall, IMEPay's sentiment is 41.8% positive, 57.7% negative, and 5% neutral. IMEPay's strength lies in its efficient interface, but persistent issues in Fulfillment, System Availability, and Privacy underscore the need for improvements to enhance user trust and experience.

The study shows clear differences in how users perceive Nepal's leading e-wallets. Efficiency was the most frequently discussed dimension, with IMEPay praised for its ease of use, while eSewa and Khalti faced criticism for issues related to KYC, login, and transactions. Fulfillment was a recurring weakness, such as delays, failed transfers, and unresolved cashback reduced trust. System availability was another major concern across all platforms, with frequent crashes and downtimes highlighting the need for stronger infrastructure. Privacy, though less frequently mentioned, attracted overwhelmingly negative feedback due to data security and KYC concerns. Overall, while e-wallets have gained rapid adoption in Nepal, unresolved issues in reliability, fulfillment, and privacy remain critical barriers to sustaining user trust and loyalty.

5 Implications and Conclusion

This study offers significant theoretical and practical contributions to understanding e-service quality in Nepal's e-wallet industry. Theoretically, it demonstrates that the e-SERVQUAL model remains applicable in emerging digital financial markets and shows how sentiment analysis of user-generated reviews can capture spontaneous, real-time consumer perceptions more effectively than traditional surveys.

Practically, the findings highlight clear priorities for service improvement. Efficiency must be strengthened through a smoother KYC process, reliable login systems, and intuitive designs. Fulfillment requires enhanced transaction reliability, transparent cashback policies, and more responsive customer support. System availability should be bolstered by investing in stronger infrastructure to minimize crashes and downtime. Privacy concerns demand clearer communication on data usage, stronger encryption, and user-centered consent mechanisms. Furthermore, the recent development of a merger agreement between Khalti and IMEPay has the potential to reshape Nepal's e-wallet landscape [17]. Their combined resources and user base may enable stronger competition with eSewa. However, the success of this consolidation will depend on whether the new entity can resolve persistent issues of fulfillment, system stability, and privacy that continue to erode user trust.

The study also presents policy implications for Nepal's evolving digital payments ecosystem. As the government and Nepal Rastra Bank advance regulatory reforms, there is an opportunity to introduce standardized service reliability benchmarks, incident-reporting protocols, and robust data protection requirements tailored to digital wallets. Strengthening oversight of KYC procedures, data retention, and third-party integrations can reduce privacy risks. Clear guidelines on dispute resolution and refund timelines will further enhance consumer protection.

This study has several limitations. Relying solely on Google Play reviews restricts generalizability, as user demographics differ across platforms. Although the machine learning classifier was validated, it may interpret some or ambiguous comments. Additionally, focusing on sentiment polarity limits deeper emotional insights. Future research should incorporate cross-platform datasets, multilingual sentiment models, and mixed-method validation, such as surveys and interviews, to strengthen the robustness of findings.

References

1. Dahlberg, T., Mallat, N., Ondrus, J., Zmijewska, A.: Past, present and future of mobile payments research: a literature review. Electron. Commer. Res. Appl. **7**(2), 165–181 (2008)
2. Firpo, J.: 'E-Money–Mobile Money–Mobile Banking–What's the Difference?'(2009). https://blogs.worldbank.org/en/psd/e-money-mobile-money-mobile-banking-what-s-the-difference. Last accessed 12 June 2025
3. Statista, Digital Payments-Worldwide (2025). https://www.statista.com/outlook/fmo/payments/digital-payments/worldwide?currency=USD. Last accessed 06 July 2025
4. Sharma D.: Top Digital Wallets in Nepal/Mobile payment Apps (2025). https://www.nepalitelecom.com/2019/03/top-mobile-wallets-online-digital-payment-providers-in-nepal.html. Accessed 25 July 2025
5. Nurjang, K., Ankit, K.: Raising good governance through digital payment systems: insights from Nepal and India's UPI. Butwal Campus J. **7**(2), 1–14 (2024)
6. Nepal Rastra Bank: 'Payments Systems Overshight Report FY 2023/24. https://www.nrb.org.np/contents/uploads/2025/01/Payment-Oversight-Report-2023-24.pdf. Last accessed 29 July 2025
7. Zeithaml, V.A.: 'Service quality, profitability, and the economic worth of customers: what we know and what we need to learn. J. Acad. Market. Sci. **28**(1), 67–85 (2000)
8. Bhandari, R.K., Koirala, A.N.: The role of digital payment systems in facilitating remittance transfers to Nepal. Law Econ. **3**(12), 15–19 (2024)
9. Parasuraman, A., Zeithaml, V.A., Malhotra, A.: ES-QUAL: a multiple-item scale for assessing electronic service quality. J. Serv. Res. **7**(3), 213–233 (2005)
10. Bing, L.: 'Sentiment analysis and opinion mining (synthesis lectures on human language technologies)', University of Illinois: Chicago. IL, USA (2012)
11. Gurung, K., Gnawali, S.: Emoji votes: predicting Kathmandu's 2022 Mayoral election via Facebook Sentiment. J. Inf. Sci. Theory Pract. **13**(3), 72–84 (2025)
12. Upreti, A., Pokhrel, L., Ban, H., Gurung, K.: 'Decoding digital campaigns: a multi-method analysis of Facebook posts during 2022 parliamentary elections in Nepal. Asia Pac. J. Inf. Syst. **35**(1), 195–215 (2025)
13. Mohammad, S.M., Dorr, B.J., Hirst, G., Turney, P.D.: Computing lexical contrast. Comput. Linguist. **39**(3), 555–590 (2013)
14. Sari, P.K., Alamsyah, A., Wibowo, S.: 'Measuring e-Commerce service quality from online customer review using sentiment analysis. Journal of Physics: Conference Series, pp. 012053, IOPScience, Indonesia (2018)
15. Adiningtyas, H., Auliani, A.S.: Sentiment analysis for mobile banking service quality measurement. Procedia Comput. Sci. **234**, 40–50 (2024)
16. Gurung, K.: 'Harnessing sentiment analysis for political campaigns: trends, platforms, and future directions. JeDEM-eJournal eDemocracy Open Gov. **17**(3), 59–77 (2025)
17. Albu, S.: Nepal's Khalti and IME Pay sign MoU for merger. https://thepaypers.com/payments/news/nepals-khalti-and-ime-pay-sign-mou-for-merger. Last accessed 20 July 2025

Reinforcement-Guided Swarm Drones: Effects of PSO and Pheromones on Collective Behaviour

Neha Gupta, Vivek Tetarwal(✉), and Amit Mathur

Central Research Laboratory, Bharat Electronics Limited, Ghaziabad, India
{neha.gupta,vivektetarwal,amitmathur}@bel.co.in

Abstract. Swarm robotics draws inspiration from natural systems to develop scalable, decentralized control strategies for multi-agent systems. In this work, we explore the integration of Reinforcement Learning (RL) with Particle Swarm Optimization (PSO) and pheromone-based mechanisms for autonomous swarm drone navigation in a grid-based environment. We evaluate and compare the impact of PSO, pheromone communication, and their combination on collective performance, convergence speed, and path efficiency. Additionally, ablation studies are conducted to examine the performance of the swarm in the absence of either or both coordination strategies. Experimental results demonstrate that combining PSO with pheromone guidance significantly enhances success rate and learning stability. Conversely, removing both mechanisms leads to a drastic drop in performance, emphasizing the importance of global and local information-sharing in swarm intelligence. These findings contribute to the design of efficient, cooperative multi-drone systems for real-world applications.

Keywords: Reinforcement Learning · Particle Swarm Optimization · Pheromones · Swarm Drone Navigation

1 Introduction

Swarm robotics has emerged as a promising paradigm in the field of autonomous systems, leveraging the principles of collective intelligence and decentralized control to achieve scalable and robust multi-agent behaviour. Inspired by the coordinated actions observed in biological systems—such as ant colonies, bird flocks, and fish schools—swarm robotics enables a group of relatively simple agents to collaboratively perform complex tasks. Such systems exhibit inherent advantages including fault tolerance, flexibility, scalability, and adaptability to dynamic environments. These properties make swarm-based strategies particularly attractive for applications such as environmental monitoring, search-and-rescue operations, agricultural surveillance, and exploration in unknown or hazardous terrains.

Despite the potential, designing effective coordination mechanisms that balance autonomy with cooperation in swarm systems remains a critical challenge. Traditional rule-based or behaviour-driven methods often lack adaptability and fail to generalize across diverse scenarios. RL, a trial-and-error based learning paradigm, offers a compelling alternative by allowing agents to learn optimal policies through interactions with

H.-S. Ma and Y.-W. Chan (Eds.): IC 2026, LNEE 1646, pp. 119–128, 2027.
https://doi.org/10.1007/978-981-92-0779-4_13

the environment. While RL has shown success in single-agent settings, extending it to multi-agent or swarm contexts introduces several difficulties, including sparse rewards, non-stationary due to changing policies of other agents, and inefficient exploration.

To mitigate these challenges, hybrid frameworks that augment RL with bio-inspired coordination strategies have gained traction. In this study, we investigate two such strategies—PSO and pheromone-based communication—and analyse their influence on the learning performance and emergent behaviour of a swarm of autonomous drones.

PSO, inspired by the collective movement of bird flocks, enables agents to share global knowledge by dynamically adjusting their trajectories based on the historically best-known positions of themselves and their peers. On the other hand, pheromone communication, modelled after the stigmergic behaviour of social insects like ants, allows agents to interact indirectly by modifying and sensing a shared environmental variable—pheromone intensity—thus facilitating local coordination and memory of explored regions.

This work presents a systematic study on how PSO and pheromones, individually and in combination, affect swarm behaviour when integrated with RL. We explore four distinct configurations: RL with both PSO and pheromones, RL with PSO only, RL with pheromones only, and RL with neither PSO nor pheromones (baseline).

We evaluate these configurations using a team of drones navigating a grid-based environment toward a predefined goal. Performance is measured through various quantitative metrics including average episode reward, episode steps, learning loss, success rate, and the spatial distribution of pheromone traces. Our findings highlight that the integration of both PSO and pheromones leads to faster convergence, more stable learning dynamics, and higher success rates. The ablation results indicate that removing either mechanism degrades performance, and the absence of both results in a collapse of coordinated behaviour. These results demonstrate the critical role of combining global optimization and local communication in enhancing the efficiency and robustness of swarm learning.

By offering a detailed comparative analysis, this work contributes to the understanding of hybrid learning-control strategies for decentralized multi-agent systems and informs future designs of intelligent swarm robotics in real-world applications.

2 Related Work

Swarm robotics, drawing inspiration from biological systems such as ant colonies and bird flocks, has emerged as a promising paradigm for coordinating large numbers of simple agents to achieve complex collective behaviours. Fundamental to this field are bio-inspired optimization techniques like Ant Colony Optimization (ACO) [1] and PSO [2], which enable decentralized decision-making and adaptive exploration through indirect communication mechanisms, such as pheromone trails or velocity alignment. In recent years, RL has significantly advanced the capabilities of individual agents in robotics by enabling them to learn policies through trial and error. The seminal works of Sutton and Barto [3] laid the theoretical foundations of reinforcement learning. Early policy-gradient methods, particularly the REINFORCE algorithm proposed by Williams [5], demonstrated that agents could directly optimize parameterized policies using stochastic gradient ascent, establishing the basis for modern policy optimization techniques. The introduction of Deep Q-Networks (DQNs) by Mnih et al. [4] marked a turning point in robotic control and policy optimization by combining deep neural networks with value-based reinforcement learning. RL's ability to handle high-dimensional state-action spaces has made it particularly appealing for autonomous navigation and path planning.

However, scaling RL to multi-agent settings introduces several new challenges, such as non-stationary environments, credit assignment, and the need for coordination. To address these, Multi-Agent Reinforcement Learning (MARL) techniques have been developed. Notable contributions in this space include centralized training with decentralized execution, parameter sharing among homogeneous agents, and communication-based learning [6, 7]. These approaches have enabled more stable and efficient learning in swarm-based tasks.

Integrating biologically inspired heuristics with learning-based approaches is an emerging trend aimed at leveraging the strengths of both paradigms. For instance, Zhang et al. [8] introduced a hybrid algorithm that couples RL with PSO to strike a balance between exploration and exploitation. In our earlier work [9], we compared PSO, ACO, and a hybrid ACO-PSO for multi-target drone search, showing that the hybrid method outperformed the stand-alone approaches by better balancing exploration and exploitation. This synergy allows agents to dynamically adapt their movement strategies based on learned experiences and swarm-wide trends.

In aerial robotics, there has been increasing interest in applying MARL to control drone swarms for tasks such as surveillance, area coverage, and target tracking. Batra et al. [10] demonstrated the effectiveness of end-to-end DRL for decentralized control of quadrotor swarms, showing improved robustness and scalability. Similarly, Li et al. [11] used graph-based MARL frameworks to enable scalable control policies in UAV swarms, which are particularly suited for dynamic, obstacle-rich environments. These systems benefit from both individual policy optimization and global coordination based on inter-agent communication. Furthermore, studies like those by Arranz et al. [12] and Tuncer et al. [13] have shown the potential of DRL in real-world applications such as ground surveillance and wildfire reconnaissance. These works underscore the importance of reliable learning frameworks for coordinating large groups of aerial agents in uncertain environments.

Despite these advancements, the literature lacks comprehensive studies evaluating the combined influence of PSO and pheromone-based mechanisms within RL frameworks. Most current works either focus on optimizing single-agent behaviours or rely purely on heuristic strategies without learning. Our work aims to fill this gap by systematically analysing how PSO and pheromone signalling affect the learning dynamics and emergent behaviours of reinforcement-guided drone swarms, thereby contributing to the design of more adaptive and efficient multi-agent systems.

3 Methodology

This study investigates the impact of PSO and pheromone-based mechanisms on the RL-based behaviour of swarm drones in a grid-based environment. The agents are trained to navigate from a starting position to a goal region, with the objective of maximizing cumulative reward while minimizing path length and learning time. The overall procedural flow, from environment setup to training and result logging, follows a structured pipeline as illustrated in Fig. 1, capturing the interactions between initialization, learning, coordination, and evaluation stages.

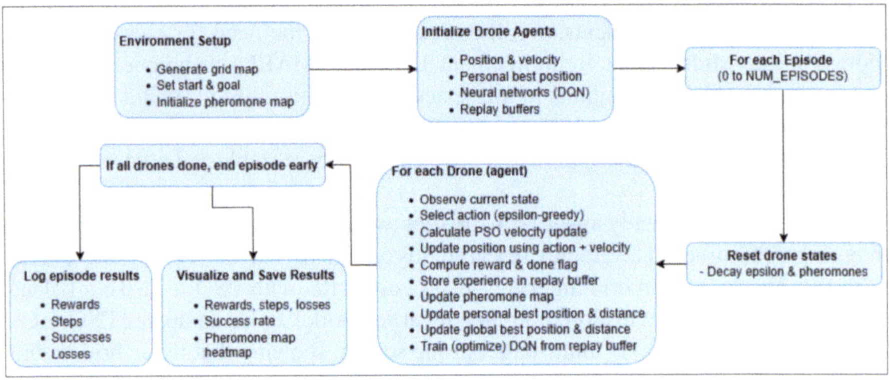

Fig. 1. Flowchart Overview

3.1 Environment, Task, Agent Model and State Representation

The simulation is conducted on a 20 × 20 2D grid world environment, where multiple autonomous drones navigate from their respective start positions to a common goal. The environment includes obstacles, and agents must learn to avoid collisions while minimizing travel distance and time. Each drone operates independently but shares the same learning policy through parameter sharing.

Each drone is modelled as an agent with a discrete action space: move in one of four cardinal directions or stay in place. A reward signal is provided at each step based on the agent's proximity to the goal and the efficiency of its path. The state representation for each agent includes: Agent's position, Relative position to the goal, Local pheromone concentration, neighboring drone locations within a sensing radius. The inclusion of pheromone and PSO components depends on the experiment configuration.

3.2 RL Framework, PSO Integration and Pheromone Communication

Each drone operates under an independent Deep Q-Network (DQN), trained with the standard Bellman update:

$$Q(s_t, a_t) := Q(s_t, a_t) + \alpha \left[r_t + \gamma . \max_{a'} Q\big(s_{\{t+1\}}, a'\big) - Q(s_t, a_t) \right] \tag{1}$$

where: $Q(s_t, a_t)$: Q-value of action a_t at state s_t; α is the learning rate; γ is the discount factor; s and a denote the current state and action; r is the reward.

Exploration is managed using an ε-greedy policy. The state includes the agent's position, goal location, and optionally, pheromone concentration in neighboring cells.

In the PSO-enhanced setting, each agent maintains a velocity vector updated using both personal and global best-known states:

$$v_i(t+1) = \omega v_i(t) + c_1 r_1 (p_i^{best} - x_i(t)) + c_2 r_2 (g^{best} - x_i(t) \quad (2)$$

$$x_i(t+1) = x_i(t) + v_i(t+1) \quad (3)$$

where: x_i is the current position; p_i^{best} the personal best; g^{best} the global best; and ω, c_1, c_2 are inertia and acceleration coefficients.

The swarm's global best information is shared to help coordinate movement direction among agents.

In the PSO-enhanced setting, each agent maintains a velocity vector updated using both personal and global best-known states. For pheromone-based coordination, each agent deposits a fixed pheromone value into its current cell. A global pheromone map is updated at each time step with evaporation dynamics modelled as:

$$P_{i,j}(t+1) = (1-\rho)P_{i,j}(t) + \Delta P_{i,j} \quad (4)$$

where: ρ is the evaporation rate and $\Delta P_{i,j}$ is the pheromone deposited.

Agents can sense local pheromone concentrations within their immediate vicinity, influencing their state representation and action selection.

3.3 Experiment Configurations

Each episode proceeds through a loop where agents observe their state, choose actions, update positions using actions (with or without PSO influence), receive rewards, update Q-values, and modify pheromone maps. These steps are iteratively executed for each drone and episode, with early termination if all agents reach the goal. Metrics such as cumulative reward, steps to completion, DQN loss, and pheromone distribution are logged after each episode.

To evaluate the contribution of each mechanism, the following configurations were tested: **RL + PSO + Pheromones** (Full integration of both coordination strategies); **RL + PSO** (Only global optimization used for coordination); **RL + Pheromones** (Only local stigmergic communication employed) and **RL Only/Baseline** (No coordination mechanisms).

Each model is trained for 200 episodes, and performance is evaluated using reward per episode, success rate, DQN loss, step efficiency, and heat maps of pheromone activity.

4 Results

This section presents the performance of the proposed drone swarm system under four configurations: **(1) Baseline (PSO + Pheromones), (2) Without PSO, (3) Without Pheromones,** and **(4) Without PSO and Pheromones.** Evaluation is based on pheromone maps, episode rewards, steps per episode, training losses, and success rates.

Figure 2 presents the ablation results across four experimental configurations: Full model (PSO + Pheromone), No PSO Velocity, No Pheromone, and No PSO & No Pheromone. While all configurations eventually achieved high success rates, the learning dynamics and efficiency varied considerably across the variants. As shown in the leftmost subplot, the Full model and No Pheromone variants reached a 100% success rate almost immediately, demonstrating rapid convergence. In contrast, the No PSO Velocity and No PSO & No Pheromone variants displayed delayed and unstable success rates, requiring over 70 episodes to stabilize, with intermittent failures persisting longer.

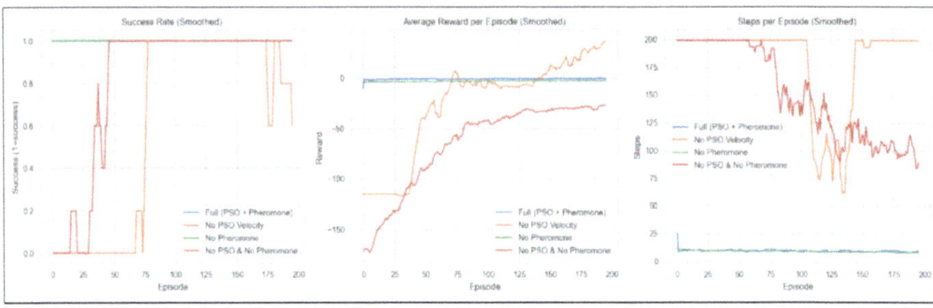

Fig. 2. Ablation Summary

4.1 Baseline: PSO+ Pheromone Mechanism

The baseline configuration integrates PSO for global coordination with a pheromone mechanism for local RL. As illustrated in Fig. 3, the final pheromone map shows clearly defined trails concentrated along optimal paths, indicating frequent drone traversal and successful reinforcement of efficient routes. This demonstrates the drones' ability to collectively exploit shared information and converge on globally optimal trajectories.

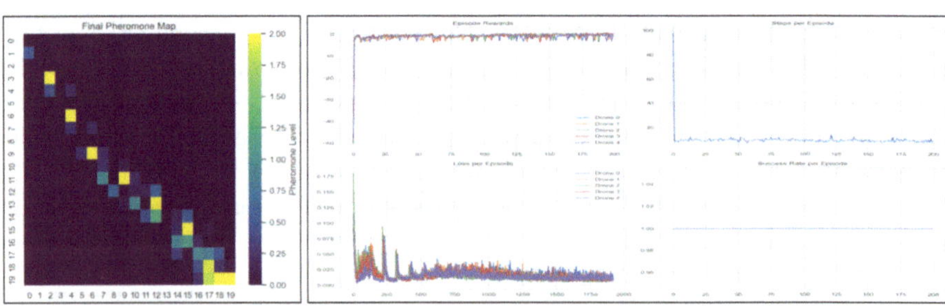

Fig. 3. Baseline Model Final Pheromone Map and Training Metrics

Training metrics in Fig. 3 highlight stable learning behaviour. Episode rewards for all drones rapidly improve within the first 50 episodes and stabilize near zero, indicating effective task completion with minimal penalties. Steps per episode decline significantly after the initial episodes and remain consistent, reflecting efficient path planning. The loss curve shows an initial spike due to exploratory actions but gradually stabilizes as policies converge. Importantly, the success rate maintains a constant 100% across all episodes, underscoring the robustness of the combined PSO and pheromone approach for coordinated multi-drone navigation.

4.2 Without PSO

In this setup, global coordination via PSO is removed, and the drones rely solely on local pheromone-based feedback. As shown in Fig. 4, the pheromone map exhibits highly concentrated trails along narrow regions, indicating that drones predominantly exploit local pheromone cues with minimal exploration. Compared to the baseline, path diversity is markedly reduced, suggesting less optimal coverage of the environment.

Training metrics in Fig. 4 highlight the limitations of removing PSO. Episode rewards improve steadily but converge more slowly, reaching a plateau around episode 150. The step count remains constant at approximately 200 steps per episode, signifying less efficient navigation. Loss values decline gradually over extended training but exhibit intermittent spikes, indicating slower stabilization of learning. Despite these inefficiencies, the success rate persists at 100%, showing that while the drones complete their tasks, they do so with reduced efficiency and slower convergence compared to the baseline.

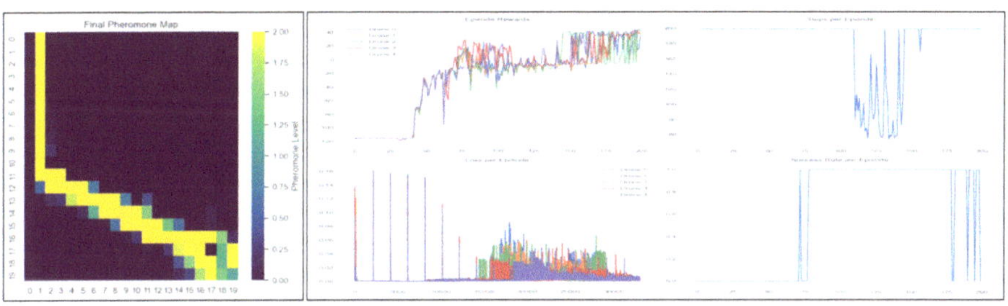

Fig. 4. No PSO Final Pheromone Map and Training Metrics

4.3 Without Pheromones

In this configuration, the pheromone mechanism is disabled, and swarm coordination relies solely on PSO. As shown in Fig. 5, the pheromone map is uniformly flat, confirming the absence of pheromone deposition and local environmental reinforcement. Coordination is entirely governed by position and velocity updates from the PSO algorithm.

Despite the removal of pheromones, Fig. 5 demonstrates that the system achieves a stable 100% success rate, with rewards steadily improving over episodes. The number of steps per episode shows a clear declining trend, indicating increasing efficiency as training progresses. Loss values also decrease consistently, stabilizing over time, which reflects effective convergence under PSO-driven control.

While performance remains strong, the absence of pheromone feedback removes fine-grained local guidance, which may limit adaptability in more dynamic or complex environments. This result emphasizes that PSO alone can drive successful goal-reaching behavior, but pheromones contribute to enhanced adaptability and local decision-making.

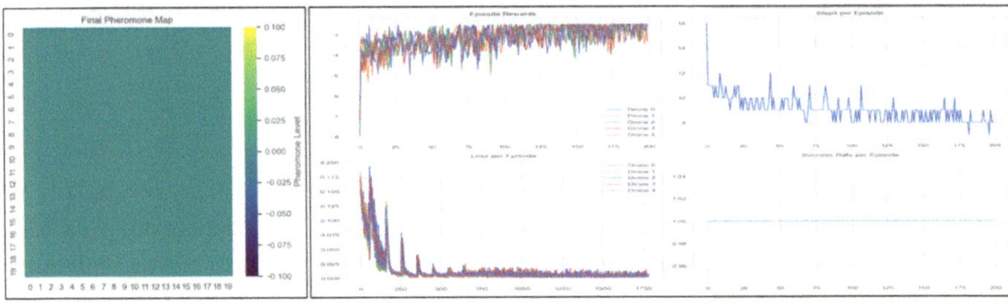

Fig. 5. No Pheromone (a) Final Pheromone Map and (b) Training Metrics

4.4 Without PSO+ Pheromones

In this ablated configuration, both PSO and the pheromone mechanism are disabled, resulting in drones operating independently without any global or local coordination. As shown in Fig. 5, the pheromone map remains uniformly flat, confirming the absence of any pheromone deposition or environmental reinforcement.

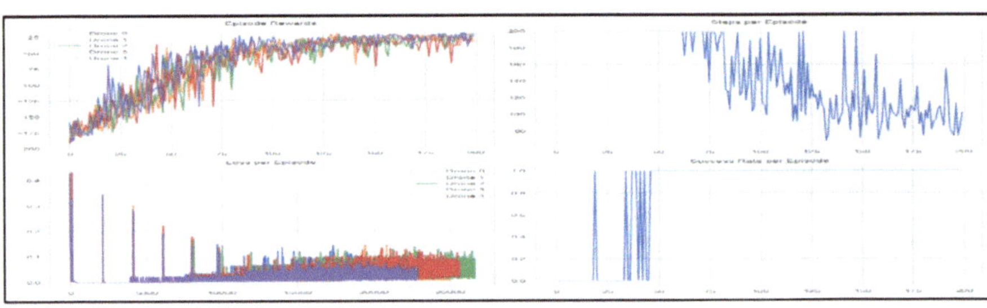

Fig. 6. No PSO No Pheromone Training Metrics

The training metrics in Fig. 6 reveal significantly degraded performance. Rewards exhibit large fluctuations early on but gradually stabilize over time. The steps per episode start at the maximum limit (200 steps) with no goal attainment but gradually decline as learning progresses, coinciding with intermittent increases in the success rate. After approximately 50 episodes, the success rate improves sharply and reaches 100%, aligning with the reduction in steps, indicating that the agents eventually learn to reach the goal more efficiently despite the lack of coordination mechanisms.

Loss per episode displays high variability across drones, reflecting instability in individual learning without shared reinforcement signals. While the agents ultimately achieve task success, the lack of both global and local coordination mechanisms leads to slower convergence and reduced learning efficiency, emphasizing the importance of coordinated strategies for scalable and robust swarm behavior.

In Table 1, the comparative analysis highlights the synergistic impact of combining PSO with pheromones for swarm coordination. The PSO+ Pheromones configuration demonstrates superior performance, achieving fast convergence, stable learning, low reward variance, and balanced exploration. Removing PSO reduces global coordination, leading to more localized but still stable behavior due to pheromone-based reinforcement. Conversely, disabling pheromones while retaining PSO results in broader but less efficient exploration, as local reinforcement cues are absent. The fully ablated configuration (No PSO, No Pheromones) suffers from slow and unstable learning, high variance in rewards, and erratic step counts, though eventual success is achieved after extended training. These findings underscore the complementary roles of PSO and pheromone mechanisms in enhancing both learning efficiency and scalability in swarm systems.

Table 1. Summary of Performance across Configurations

Configuration	Path Optimization	Learning Stability	Reward Variance	Success Rate	Exploration
PSO+ Pheromones	High	High	Low	100%	Balanced
No PSO	Moderate	Moderate	Medium	100%	Localized
No Pheromones	Low	Moderate	Medium	100%	Broader but Inefficient
No PSO, No Pheromones	None	Low	High	Initially Variable → 100%	Poor and Slow

5 Conclusion

In this work, we presented a comprehensive analysis of swarm-based coordination strategies using reinforcement learning for multi-agent drone navigation in a grid environment. By integrating Particle Swarm Optimization (PSO) and pheromone-based communication into a Deep Q-Network framework, we investigated their individual and combined impacts on collective drone behaviour.

Our experimental results demonstrate that the combined use of PSO and pheromone cues significantly enhances learning stability, reward accumulation, and task efficiency compared to using either method in isolation or relying solely on reinforcement learning. The PSO mechanism facilitates faster convergence by guiding agents toward historically optimal paths, while the pheromone map enables local, decentralized decision-making that reinforces successful trajectories.

In contrast, agents operating without swarm-inspired strategies exhibit slower learning and less coordinated behaviour. The absence of pheromone cues results in a lack of environmental memory, and the removal of PSO coordination leads to fragmented exploration and reduced path efficiency.

These findings validate the synergistic benefits of bio-inspired techniques in swarm intelligence and highlight the importance of combining global guidance (via PSO) with local stigmergy (via pheromones) in cooperative robotic systems. Future work may extend this framework to dynamic environments, 3D navigation, and real-world deployment of autonomous drone fleets in search-and-rescue or surveillance missions.

References

1. Dorigo, M., Stützle, T.: Ant Colony Optimization. MIT Press, Cambridge (2004)
2. Kennedy, J., Eberhart, R.C.: Particle swarm optimization. In: Proc. IEEE Int. Conf. Neural Netw., Perth, Australia, pp. 1942–1948 (1995)
3. Sutton, R.S., Barto, A.G.: Reinforcement Learning: An Introduction, 2nd edn. MIT Press, Cambridge (2018)
4. Mnih, V., et al.: Human-level control through deep reinforcement learning. Nature **518**, 529–533 (2015)
5. Williams, R.J.: Simple statistical gradient-following algorithms for connectionist reinforcement learning. Mach. Learn. **8**(3–4), 229–256 (1992)
6. Foerster, Y., et al.: Stabilising experience replay for deep multi-agent reinforcement learning. In: Proc. 34th Int. Conf. Mach. Learn., pp. 1146–1155 (2017)
7. Foerster, J.N., et al.: Learning to communicate with deep multi-agent reinforcement learning. In: Advances in Neural Information Processing Systems, pp. 2137–2145 (2016)
8. Zhang, Y., et al.: A reinforcement learning-based particle swarm optimization algorithm for improved symmetry between convergence speed and diversity. Symmetry **16**(10), 1290 (2024)
9. Gupta, N., Mathur, A.: Performance comparison of PSO, ACO and hybrid ACO-PSO for multi-target search using autonomous swarm drones in obstacle-rich environments. In: Proc. IEMENTech 2025, IEEE (2025)
10. Batra, S. et al.: Decentralized control of quadrotor swarms with end-to-end deep reinforcement learning. In: Proc. 5th Conf. Robot Learn., pp. 576–586 (2022)
11. Li, Y., et al.: Graph-based multi-agent reinforcement learning for large-scale UAV swarm control. Aerospace Sci. Technol. **150**, 109166 (2024)
12. Arranz, R., et al.: Application of deep reinforcement learning to UAV swarming for ground surveillance. Sensors **23**(21), 8766 (2023)
13. Tuncer, E., et al.: A deep reinforcement learning algorithm for trajectory planning of swarm UAVs for wildfire reconnaissance. Electronics **13**(13), 2568 (2024)

Author Index

H.-S. Ma and Y.-W. Chan (Eds.): IC 2026, LNEE 1646, p. 129, 2027.
https://doi.org/10.1007/978-981-92-0779-4

MIX
Papier aus verantwortungsvollen Quellen
Paper from responsible sources
FSC® C105338

FSC
www.fsc.org

If you have any concerns about our products,
you can contact us on
ProductSafety@springernature.com

In case Publisher is established outside the EU,
the EU authorized representative is:
Springer Nature Customer Service Center GmbH
Europaplatz 3, 69115 Heidelberg, Germany

Printed by Libri Plureos GmbH
in Hamburg, Germany